Linear Integrated Circuits

Linear Integrated Circuits

Joseph J. Carr

Butterworth-Heinemann
Linacre House, Jordan Hill, Oxford OX2 8DP
A division of Reed Educational and Professional Publishing Ltd

 A member of the Reed Elsevier plc group

OXFORD   BOSTON   JOHANNESBURG
MELBOURNE   NEW DELHI   SINGAPORE

First published 1996

British Library Cataloguing in Publication Data
A catalogue record for this book is available from the British Library

ISBN 0 7506 2591 0

Library of Congress Cataloguing in Publication Data
A catalogue record for this book is available from the Library of Congress

Printed and bound in Great Britain by Hartnolls Limited, Bodmin, Cornwall
Typeset by Laser Words, India

Contents

Preface

The field of linear integrated circuit (IC) electronics exploded onto the scene in the 1960s. Early offerings, such as the mA-703 RF/IF amplifier and the mA-709 operational amplifier, were terribly expensive. Offerings that cost less than £1 today, and are used only for replacement or by hobbyists, cost £50 in those days. Because those devices offered some tremendous technical advantages, they were often used despite the price. Partly because of the well-known 'learning curve' in the semiconductor industry, and partly because of improving technology, the prices of linear integrated circuits fell rapidly. Today, even the costliest special-purpose linear IC devices cost only a fraction of what the limited capability mA-709 cost in the mid-1960s.

Not only has cost dropped, but performance and what is loosely called 'functionality' has increased dramatically. The operational amplifier of 1970 choked if asked to process signal frequencies above 25 KHz or so (and even less for 'frequency compensated' devices), while op-amps are available today with gain-bandwidth products of 500 and 1000 MHz. The input impedance of early devices was limited to a few megohms, while today we can buy devices with terraohm input impedances. Even in an era of the digitalization of electronics, the analogue IC device still occupies an important role in circuit design, especially in data acquisition, instrumentation and control circuits.

This textbook is intended for both classroom use and independent study. Throughout the text are objectives for each chapter, pre-quiz questions to be answered prior to reading the chapter, a summary and then questions and problems.

Joseph J. Carr, B.Sc., MSEE

Free supplemental software offer

Accompanying this book is a suite of supplemental software written by the author to assist in the design of linear IC circuits. A collection of circuits are offered covering a range of applications. For each circuit a simple slider bar operation allows the user to alter the parameters until a desired combination is achieved. Topics covered include inverting and noninverting amplifiers, differential amplifiers, instrumentation amplifiers, waveform generators and filters. This software is available for downloading from Butterworth-Heinemann's website:

http://www.butterworth.heinemann.co.uk/carr/carr.html

It is also available on disk for a small handling charge. Please contact Duncan Enright, Newnes Publisher, at Butterworth-Heinemann:

e-mail: duncan.enright@bhein.rel.co.uk
(make the subject of your message 'Carr Software Offer')

or write to:

Duncan Enright, Newnes Publisher
Butterworth-Heinemann
Linacre House, Jordan Hill
Oxford OX2 8DP, UK.

CHAPTER 1

Introduction to linear integrated circuit devices

OBJECTIVES

1. Learn the advantages of IC devices compared with discrete circuitry.
2. Be able to identify the different IC package types.
3. Learn the differences between linear and digital devices.
4. Understand the power requirements and methods of construction used with linear IC devices.

1-1 PRE-QUIZ

These questions test your prior knowledge of the material in this chapter. Try answering them before you read the chapter. Look for the answers (especially those you answered incorrectly) as you read the text. After you have finished studying the chapter try answering these questions again, and those at the end of the chapter (see Section 1-12).

1. List four common types of IC package.
2. A(n) _____ signal is continuous in both range and domain, and has a transfer function of the form $V = F(t)$.
3. _____ amplifiers normally operate with bipolar DC power supplies, even though no ground terminal exists on the device.
4. The transfer function of an operational transconductance amplifier is of the form:

   _____ .

1-2 A BRIEF HISTORY OF INTEGRATED CIRCUITS _____

The modern era of 'solid-state' electronics began in the late 1940s with the invention of the bipolar transistor. Only a decade later engineers were working to build a device containing multiple transistors and resistors formed onto a single semiconductor substrate. By the early 1960s the 'integrated circuit' (IC) was born at Fairchild in California, and the modern chip revolution began. Today, hundreds of millions of IC devices are sold annually, and the IC semiconductor industry is a major contributor to the world economy.

Two of the earliest commercially successful IC devices were the μA-703 RF/IF gain block and the μA-709 operational amplifier. The μA-703 device was a simple high frequency gain block, and was frequently used as such in the mid-1960s. These devices were frequently used as the FM intermediate frequency (IF) amplifier in broadcast and communications receivers. Typically, three to five μA-703 devices were used to provide the 60 to 80 dB gain normally required of an IF amplifier in consumer FM broadcast receivers.

The μA-709 was an operational amplifier, or 'op-amp'. Invented in 1948 (in vacuum tube form) by George Philbrick, the operational amplifier gets its name from the fact that it was originally intended to perform *mathematical operations* in analog computers. The unique property of the op-amp that makes it so useful and so versatile is that the overall circuit transfer function (e.g. V_o/V_{in} for voltage amplifiers) is controlled by the feedback network between output and input. Although perhaps not immediately apparent, this attribute of the operational amplifier makes it one of the most powerful and flexible IC devices on the market.

There is a well-known production 'learning curve' in the semiconductor industry under which costs drop dramatically after an initial high-priced period. In 1964, for example, electronic distributor catalogs in the USA listed the μA-709 operational amplifier for $110.00, and two years later it was still $79.00; today the μA-709 costs less than $1.00 where it is still available at all. Devices with greatly superior performance compared with the μA-709 now cost only a few dollars and are easily available. Production yields and device availability is considerably improved over the early years. Where designers in 1963 had to wait for 'rationed' parts, operational amplifiers are available today from distributor stock inventories in almost any quantity that may be needed. Even electronic hobbiests can buy common operational amplifiers in bulk at retail stores for very small amounts of money.

Integrated circuit electronics became so successful so quickly because it allows immense circuit density, permitting many circuits to be implemented with fewer overall discrete components. A very large number of devices can be packaged in a very small volume. Figure 1-1A shows a photograph of an integrated circuit, while Fig. 1-1B shows the circuitry contained within a common IC operational amplifier.

Performance can be dramatically better in IC devices compared with the discrete component version of the same circuit. Consider thermal drift of

(A)

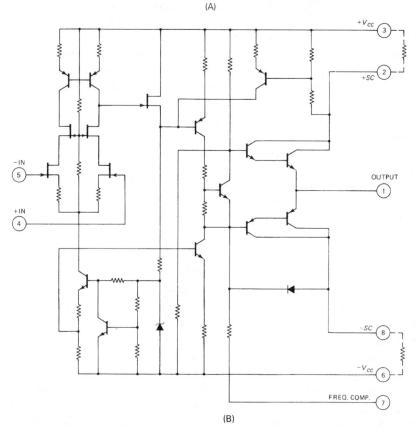

(B)

FIGURE 1-1 (A) Integrated circuit dual in-line package; (B) typical internal circuitry of an operational amplifier.

DC amplifiers, for example. All DC amplifiers suffer an undesired change in output DC offset voltage that is a function of temperature: $\Delta V_{off} = F(T)$. Even low-cost modern IC operational amplifiers are several orders of magnitude better with respect to drift than traditional discrete transistor models because all of the semiconductors and internal resistors (the main sources of drift) share the same thermal environment of a common substrate in IC devices. In discrete circuits, on the other hand, those components are spread out over several square inches of a printed circuit board and thus do not share the same thermal environment — so drift will be more pronounced than in the IC version of the same basic circuit.

Also consider issues such as stray capacitance, inductance and inadvertent series resistances due to conductor placement and lengths. These 'stray' parameters can seriously disrupt circuit operation, but are less of a factor in IC devices compared with discrete circuits because of the small sizes and short distances involved in IC design.

Cost is another great advantage of the integrated circuit. Early transistor operational amplifiers were not only larger and ran hotter than their modern IC counterparts, but they were more costly as well. In addition, those early amplifiers only poorly approximated the performance of the ideal textbook version of the amplifier (see Chapter 3). Modern IC operational amplifiers come very close to the ideal, especially as regards input impedance and open-loop gain.

Most of the spectacular benefits of modern electronics derive from integrated circuit electronics technology. From consumer devices, to commercial instruments, to military equipment, the IC either improves the product or makes it possible in the first place. Many of the marvels of modern electronics simply could not exist without the integrated circuit. This book is designed to give you an overview of the field, as well as detailed technical information on the workings and applications of the most commonly available devices.

1-3 INTEGRATED CIRCUIT SYMBOLS _____

Manufacturers and users of modern integrated electronics devices have adopted several standard circuit symbols for use in schematic diagrams. Figure 1-2 shows the most common versions of these symbols as they apply to linear IC devices.

The standard single-ended amplifier symbol is shown in Fig. 1-2A. In this type of amplifier the input signal is applied between the single input terminal and common ('earth' or 'ground' if less rigorous terminology is allowed). The symbol shown in Fig. 1-2A has a $V+$ DC power supply terminal and a ground or common terminal. In some actual schematic circuit diagrams these terminals are not shown for sake of simplicity. Do not assume, however, that they are not used in such cases.

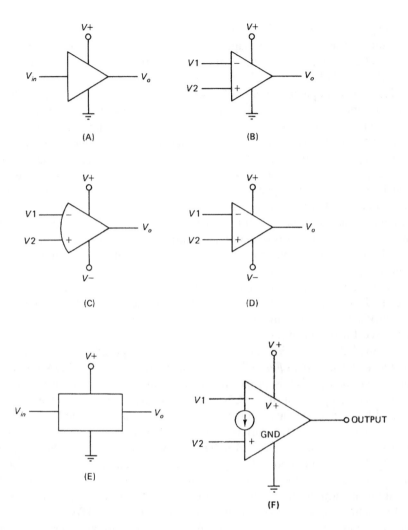

FIGURE 1-2 Circuit symbols used for linear ICs: (A) amplifier with single polarity DC supply; (B) differential amplifier; (C) operational amplifier; (D) alternative operational amplifier; (E) gain or function block; (F) current difference amplifier.

A *differential amplifier* symbol is shown in Fig. 1-2B. This type of amplifier produces an output signal that is proportional to the difference between the two input signals. If A_v is the voltage gain of the amplifier, and $V1$ and $V2$ are the two input signals, then the output voltage is defined by $A_v(V2 - V1)$.

There are always two inputs on a differential amplifier. The *inverting input* $(-IN)$ produces an output signal that is 180 degrees out of phase with the input signal. The *noninverting input* $(+IN)$ produces an output that is in-phase with the input signal.

Figures 1-2C and 1-2D show two alternative symbols used for the operational amplifier. The supposedly most proper version is that of Fig. 1-2C, although by common industry practice the more generic linear amplifier symbol of Fig. 1-2D is the *de facto* standard.

The standard operational amplifier symbol has several features. First, there are two inputs (inverting and noninverting) because most op-amps are differential amplifier devices. In fact, only the differential amplifier gives true, full-range, operational amplifier performance. Second, there are two DC power supply terminals. The '$V+$' terminal requires a potential that is positive with respect to common, while the '$V-$' terminal requires a potential that is negative with respect to common. Third, the output is single-ended, so output signal is taken between the output terminal and common.

Notice what is missing on this symbol? There is no 'ground' or 'common' terminal on the standard operational amplifier. The common connection is established by connecting together the 'cold' ends of the $V-$ and $V+$ DC power supplies (see Section 1-8).

Some manufacturers use the universal or generic IC symbol shown in Fig. 1-2E for their products. Although these devices tend to be special function products, or limited use amplifiers, the same generic symbol is sometimes also used for other purposes.

Figure 1-2F is the symbol used for a special type of amplifier called a *current difference amplifier* (CDA), or *Norton Amplifier*. The CDA produces an *output voltage* that is proportional to the difference between two *input currents*. The operation of the CDA is not exactly analogous to the op-amp (i.e. with the input voltages replaced by input currents), but is at least similar. The symbol for the CDA differs from the normal op-amp symbol in order to distinguish its unique operation. The CDA symbol shown in Fig. 1-2F is the regular differential amplifier symbol with a current source symbol added along one edge to let the reader know that current mode operation is intended.

Another form of linear IC amplifier, different from either op-amp or CDA, is the *operational transconductance amplifier*, or OTA. This type of amplifier has a transfer function that relates output current to input voltage ($\Delta I_o / V_{in}$). Since the transfer function expression has the units amperes/volts (or sub-units thereof), the transfer function 'gain' can be expressed in the units of conductance (siemens, S, is the modern unit, but in earlier texts mhos — 'ohms' spelled backwards — was used; 1 S = 1 mho). Since these are units of conductance we call the amplifier a 'transconductance amplifier'. The name 'operational' conveys the idea that some of the functions are similar to those of the operational amplifier. The OTA symbol is the same as the op-amp shown in Fig. 1-2D (sometimes with the letters 'OTA' superimposed inside the triangle).

Although there is some variation in schematic symbols by some manufacturers, the versions shown in Fig. 1-2 represent the vast majority of modern IC device applications.

1-4 LINEAR VERSUS DIGITAL IC DEVICES ————————————

The integrated circuit revolution brought us both linear and digital devices. It is, therefore, appropriate for us to consider the differences between the two types of device, especially because there are digital applications of linear devices, and linear applications of digital devices.

1-4.1 Analog versus digital signals

It is instructive to examine the differences between analog and digital signals. Although the formal mathematical difference is more rigorously defined, the diagram of Fig. 1-3 shows the concept intuitively.

An analog signal (Fig. 1-3A) is one that is continuous in both range and domain.

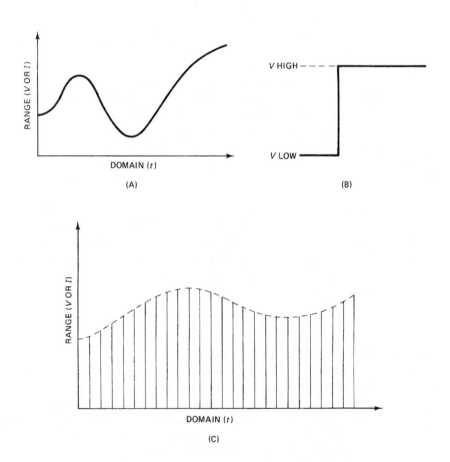

FIGURE 1-3   (A) Analog signal; (B) digital signal; (C) sampled signal.

The range of an analog signal will be either a current (I) or voltage (V), while the domain is typically time (t). Thus, an analog signal is a continuous function of the form $V = F(t)$. The actual waveshape of the analog signal is not significant for sake of this definition, even though it is often critical for applications purposes.

A digital signal is discrete (i.e. non-continuous) in range and domain.

In a digital signal the range is a voltage or current that can take on only specified values (called V_{LOW} and V_{HIGH} in Fig. 1-3B). Although a digital signal may sometimes be continuous in domain (t), most digital signals are discrete in both domain and range. The key identifying feature of the digital signal is the fact that the amplitude (range) can only take on specific discrete values. The overwhelming majority of digital signals are *binary* signals in which there are only two permissible states (e.g. V_{LOW} and V_{HIGH} in Fig. 1-3B). Modern digital computers use this type of signal.

The actual values of V_{LOW} and V_{HIGH} depends on the particular family of devices selected. In *transistor transistor logic* (TTL) devices V_{LOW} nominally ranges from zero to 0.80 volts, and V_{HIGH} is nominally +5 volts (tolerance range is +2.4 to +5.2 volts).

The low power consumption requirements of modern portable electronic devices, most notably the laptop computer, has created a demand for a 3.3 volt line of IC devices. The reduction from +5.0 volts to +3.3 volts in digital logic devices saves considerable power because the power reduction is proportional to the *square* of the voltage reduction. Although the +3.3 volt regime was originally intended for digital IC devices, the +3.3 volt power supply standard is now also used extensively in newer lines of linear IC devices.

In *complementary symmetry metal oxide semiconductor* (CMOS) devices V_{LOW} can be any voltage from zero to -15 volts, while V_{HIGH} can be any voltage from zero to +15 volts. A constraint in the CMOS system is that the condition $V_{LOW} = V_{HIGH} =$ zero is not permitted. The transition point from V_{LOW} to V_{HIGH}, or from V_{HIGH} to V_{LOW}, is one-half the difference between the two voltage levels.

A third category of signal is the *sampled analog signal* shown in Fig. 1-3C. This type of signal is *continuous in range, but discrete in domain*. The sampled signal is found extensively in instrumentation applications, especially in computerized (i.e. microprocessor based) analog instruments.

A 'linear' electrical circuit is one that shows a simple scaler relationship between output and input signals. For example, a gain of ten analog amplifier is 'linear' because the output has the same waveshape as the input, but is ten times larger. In contrast, both input and output signals of a digital circuit can take on only one of two values (V_{LOW} or V_{HIGH}); these are 'nonlinear' circuits. For purposes of our discussion, a linear IC device is one that processes analog signals (some linear circuits may also process sampled signals as well as analog signals). Also included in this admittedly loose definition are certain nonlinear circuits discussed in Chapter 10.

1-4.2 Applications of linear versus digital devices

The inherent differences between linear and digital IC devices makes most applications clear-cut and unambiguous. There are areas, however, where the situation is not so easily determined. For example, some CMOS digital gates can be biased in a way that makes them operate as linear amplifiers from DC up to AC frequencies exceeding 5 MHz. Similarly, operational amplifiers can work as clock oscillators in digital circuits, comparators, and coincidence detectors (i.e. AND gates) in a manner that suggests the binary situation that is normally expected in digital circuits. While it is true that the linear device output voltage must match the system voltage levels (V_{LOW} and V_{HIGH}) of the particular digital logic family that it works with, it is not usually difficult to interface such devices.

We will also see cases where an inherently linear IC device (such as the operational amplifier) is used to produce some output signals that are decidedly nonlinear in that the output waveshape in no way resembles the input waveshape. Examples of this type of circuit are integrators, differentiators, and logarithmic amplifiers. These are, nonetheless, considered to be applications of linear IC devices in modern terminology.

1-5 COMMON PACKAGE TYPES

The integrated circuit is formed on a tiny 'chip' of silicon material by a photolithographic process. Typical chip 'die' sizes range around 100 mil (0.100 inch), with some being larger and others being smaller. The die is typically mounted inside of a package and connected to the package pins by fine wires.

Figure 1-4 shows a die with wire attached. The connecting wires between package pins and 'solder' pads on the die are around 10 mil (0.010 inch) diameter, and are made of either gold or aluminum in most cases. Either an electric current or a thermosonic process is used to melt the end of the wire onto (and bond it with) the connecting pad on the semiconductor die.

The particular package style selected for any given IC device depends in part on the intended application and the number of pins required. For many IC devices several different packages are available. The earliest IC packages were the 6, 8, 10 and 12-lead metal can devices (Fig. 1-5A). These packages were redesigns of (and similar to) the TO-5 metal transistor package.

When viewed from the bottom of the package, the keyway marks the highest number lead or 'pin' (e.g. pin no. 8 in Fig. 1-5A), and pin no. 1 is the next pin clockwise from the keyway. One must be careful when looking at IC base diagrams to know whether a top or bottom view is depicted.

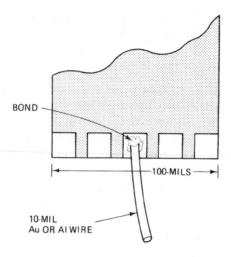

BOND

|←——————————————————100-MILS——→|

10-MIL
Au OR Al WIRE

FIGURE 1-4   Connection of gold or aluminum wire to chip die.

Perhaps the largest number of IC devices on the market today are sold in *dual in-line packages* (DIP), examples of which are shown in Fig. 1-5B. DIP packs are available in a wide variety of sizes from four to more than forty-eight pins. Although many devices are available in other size DIP packs, most linear devices are found in 8, 14 or 16-pin DIP packs.

The DIP pack is symmetrical with respect to pin count regardless of the package size, so some other means is needed to designate pin no. 1; Figure 1-5B shows several common methods. In all cases the IC DIP pack is viewed from the top. In some devices a paint dot, 'pimple', or 'dimple' will mark pin no. 1. In other cases, a semicircular or square notch marks the end where pin no. 1 is located. When viewed from the top, with the notch pointed away from you, pin no. 1 is to the left of the notch while the highest number pin is to the right of the notch.

Both plastic and ceramic materials are used in DIP pack construction. In general, the plastic packages are used in consumer and noncritical commercial (or industrial) equipment, while the ceramic packs are used in military and critical commercial equipment. The principal difference between the plastic and ceramic packages is the intended temperature range. Although exceptions exist, typical temperature range specifications are 0 to +70 degrees celcius (°C) for commercial plastic devices, and −10 to +80°C for ceramic. Military and some critical commercial ceramic devices are rated from −55 to +125°C. The principal difference between military chips and the highest grade commercial or industrial chips is the amount of testing, 'burn-in' and documentation that accompanies each device. A −55 to +125°C military device is nothing more than a −10 to +80°C commercial device that has been tested to the extended military temperature range (and tested to other

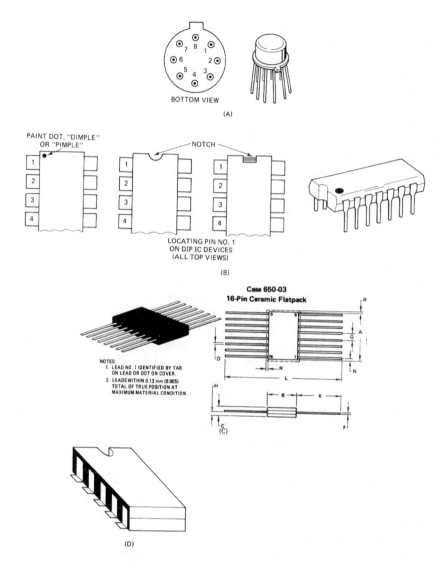

FIGURE 1-5 Integrated circuit packages: (A) metal package; (B) several varieties of dual in-line package; (C) flat-pack; (D) surface mount package.

parameters as well). Because military devices are selected from the pool of commercial devices, some people say the military ICs are nothing more than civilian devices in uniform.

An example of an IC flat-pack is shown in Fig. 1-5C. This type of package is typically used where very high component density is required. Most flat-pack devices are digital ICs, although a few linear devices are also offered. DIP package IC devices are mounted either in sockets, or by insertion of the pins through holes drilled in a printed circuit board (PCB). Flat-packs, on the other hand, are mounted on the surface by direct soldering to the conductive track of the PCB.

A relatively new (but increasingly popular) style of IC package is the *surface mounted device* (SMD), an example of which is shown in Fig. 1-5D. The SMD technology represents a significant improvement in packaging density. SMD components can be mounted closer together than other types of package, and are more amenable to automatic PCB production methods. It is expected that the SMD package will soon overtake and replace other forms, especially in Very Large Scale Integration (see below) applications.

1-6 SCALES OF INTEGRATION

There are several different scales of integration found among IC devices. The ordinary *Small Scale Integration* (SSI) device consists of single gates, small amplifiers and other smaller circuits. The number of components on each chip is on the order of twenty of less. *Medium Scale Integration* (MSI) devices have a slightly higher degree of complexity, and may have about 100 or so components on the chip. Devices such as operational amplifiers, shift registers, counters and so forth are usually classed as MSI devices. *Large Scale Integration* (LSI) devices are mostly digital ICs and include functions such as calculators, microprocessors, etc. Typical LSI devices contain from about 100 to 1000 components. Some newer devices are called *Very Large Scale Integration* (VLSI) and include some of the latest computer chips. The numbers and descriptions listed for SSI, MSI and LSI devices are approximate only, but serve to provide guidelines. Most linear IC devices are either SSI or MSI, with the latter predominating.

1-7 DISCRETE VERSUS MONOLITHIC IC VERSUS HYBRID CIRCUITS

The generic term 'microelectronics' defines two distinctly different categories of solid-state devices: *monolithic integrated circuits* (which class covers most devices discussed in this text), and *hybrid circuits*. A valid question is: 'How are monolithic and hybrid circuits different from each other, and how are they both different from regular discrete solid-state circuits?'

The monolithic IC device is made using photolithographic and other processes on a single crystal of semiconductor material. The word 'monolithic' applies because only a single piece of semiconductor material is used in the fabrication of the device (see Fig. 1-6A). A hybrid circuit, on the other hand, is a cross between monolithic and printed circuit methods. Although larger than most monolithic devices, the tight packaging makes the hybrid very unlike the printed circuit board that it superficially resembles internally. The 'macro-view' of the hybrid makes it usable as if it were an integrated circuit, even though internally the structure is quite different from IC devices.

The 'substrate diode' shown in Fig. 1-6A is formed by the interface between the P-type substrate and the regions of the semiconductor crystal that are used for the other components. This diode is a natural (if undesired) feature of IC devices. The substrate diode is normally reverse biased. If circuit conditions force the diode into the forward biased condition, then damage to the device may result. Although some applications circuit designers have been clever at using the substrate diode in certain applications, it is normally treated as if it were not there.

The typical hybrid consists of a multi-layer (usually ceramic) substrate on which conductive tracks are printed by any of several deposition methods. The integrated circuits, transistors and other semiconductor devices used in the hybrid circuit are either cemented or otherwise bonded to the ceramic substrate (Fig. 1-6B). A critical difference between the hybrid and other

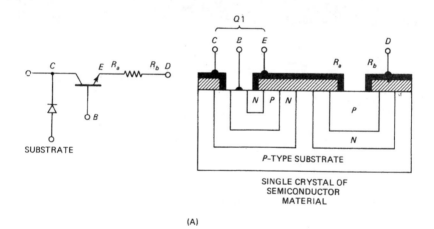

(A)

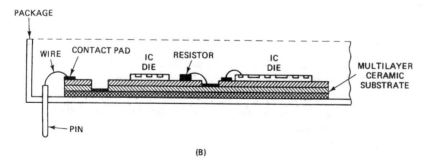

(B)

FIGURE 1-6 (A) Typical IC circuit and chip structure; (B) hybrid device; (C) examples of hybrids; (D) discrete circuit and printed circuit layout for it.

(C)

(D)

FIGURE 1-6 (continued)

'printed circuit' devices is that unpackaged semiconductors are used. The chips diodes and transistors are used in 'die' form. In other words, the unpackaged dies are installed directly onto the ceramic substrate of the hybrid. Tiny (10 mil) gold or aluminum wires connect the die electrical contacts either to the conductors of the hybrid substrate, or to the hybrid package pinouts where appropriate.

The hybrid circuit may contain a variety of components that are either difficult, expensive, or impossible to implement in monolithic IC form. Once the components are installed and the circuit is tested, the package lid is attached and the device is sealed, evacuated and prepared for final testing and shipment. Once completed, the hybrid can be treated as if it were a large integrated circuit. A photo of a several hybrids, two of them 'de-lidded', is shown in Fig. 1-6C.

A principal advantage of the hybrid is that it provides nearly IC-like packaging densities, while being relatively easy to design and manufacture without a semiconductor foundry. The hybrid is used when either circuit complexities or small quantities make monolithic IC implementation difficult or uneconomical.

Discrete electronic circuits are formed on printed circuit boards (PCB) of individually packaged active solid-state and passive components that are interconnected through photoetched copper tracks. The simple discrete circuit shown in Fig. 1-6D would easily fit onto a 100 mil (or less) IC die, or a 250 mil hybrid substrate. At normal discrete component packaging densities this same circuit would require on the order of 1 inch (1000 mil) square of area on the PCB. By comparing the ratios of $(1000 \text{ mil})^2$ to $(250 \text{ mil})^2$ and $(100 \text{ mil})^2$ we can see how IC and hybrid construction offers greatly reduced size.

1-8 LINEAR IC POWER SUPPLY REQUIREMENTS _____

Operational amplifiers and other linear IC devices normally operate from bipolar DC power supplies, such as shown in Fig. 1-7 (the pin numbers in this figure are for the so-called 'industry standard' type 741 operational amplifier). This circuit (Fig. 1-7) shows that the two DC power supplies are completely independent of each other. The $V+$ power supply is positive with respect to common, while the $V-$ supply is negative with respect to the

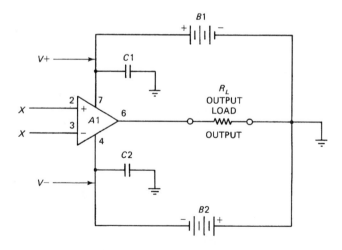

FIGURE 1-7 DC power supply arrangement for operational amplifiers.

common. The operational amplifier manufacturer will specify minimum and maximum values for $V-$ and $V+$. Typically, the maximum voltages will be on the order of ±18 volts, with some able to handle up to ±22 volts, and in at least one case ±40 volts.

Single DC Supply Operation. The operational amplifier and many other linear IC devices are designed for operation from 'split', dual or bipolar power supplies (all three terms describe the same situation). There are, however, many applications where only a single polarity DC power supply is available. In order to operate the op-amp in these cases we must either somehow supply the missing potential (e.g. with an on-board DC-to-DC converter), or devise a method for getting around the need for the missing potential.

It is reasonably easy to supply a missing potential. All that is needed is a DC-to-DC converter circuit that provides the needed voltage from the existing DC supply voltage. There are quite a few devices on the market that will produce either -15 volts from $+15$ volts (or $+12$ volts, as the case may be), or will produce isolated ±15 volt potentials from an existing non-isolated $+12$ to $+15$ volt potential.

Another method for using a single DC power supply is shown in Fig. 1-8. A resistor voltage divider, $R1/R2$, is used to bias the noninverting input of the operational amplifier to some potential between ground and $V+$; the $V-$ terminal of the device is grounded. The bias voltage on the noninverting input also appears on the output terminal as a DC offset potential. Unless the circuit following the amplifier somehow doesn't care about this offset potential, the output terminal must be capacitor coupled. The value of the

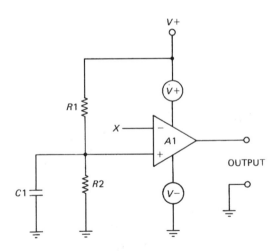

FIGURE 1-8 Single-supply operation in which the $V-$ terminal of the op-amp is grounded, and a bias network ($R1/R2$) is used to bias the noninverting input.

bias voltage is found from:

$$V1 = \frac{(V+)R2}{R1 + R2} \tag{1-1}$$

The capacitor shown in Fig. 1-8 is used to place the noninverting input at or near ground potential for AC signals, while retaining the DC level produced by the resistor voltage divider. This capacitor sometimes causes noisy operation of the device when significant 'ground loop' or 'ground plane' noise is present, so it is often omitted in practical circuits. The value of the capacitor is such that it has a capacitive reactance of less than $R2/10$ at the lowest frequency of operation. For example, if the amplifier is designed to operate down to a frequency of 10 Hz, and the value of $R2$ is 2200 ohms (a typical value in real circuits), then the value of $C1$ must be such that it has a reactance of 220 ohms or less at 10 Hz. This requirement evaluates to:

$$C1_{\mu F} = \frac{1\,000\,000}{2\pi F_{Hz} X_c}$$

$$C1_{\mu F} = \frac{1\,000\,000}{(2)(3.14)(10\ Hz)(220\ \Omega)} = 73\ \mu F$$

Because 100 μF is the next higher standard value capacitor, most designers will select 100 μF for $C1$ instead of 73 μF.

1-8.1 Protecting linear IC amplifiers

Operational and other linear IC amplifiers are sensitive to problems on the DC power supply lines. For example, the amplifier may oscillate if the DC lines are not properly decoupled. Furthermore, voltage variations, noise, and transients coupled to the DC power supply lines in one stage can affect the other stages in the same circuit, especially if power supply rejection is poor on the particular device.

We also find another problem, especially when breadboarding, trouble-shooting, and also in using portable (battery operated) equipment: reversed polarity DC power supplies. The results can be catastrophic if the DC power supply potentials are reversed. An operational amplifier with reversed DC power supplies will probably be destroyed instantly.

There are certain remedies available for each of the defects discussed above.

The problems of noise and oscillation due to cross-coupling between stages can cured by using decoupling capacitors on the amplifier power supply terminals. Capacitors $C1$ and $C2$ (each 0.1 μF) in Fig. 1-9A are used to decouple high frequencies, while the low frequency decoupling is provided by $C3$ and $C4$ (each typically 1 μF or higher). Why are two forms of capacitor sometimes needed at each op-amp power supply terminal? The higher value capacitors ($C3$ and $C4$) are typically aluminum or tantalum electrolytics. The

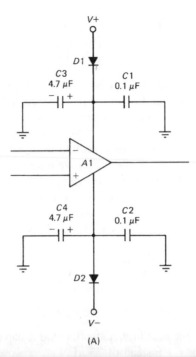

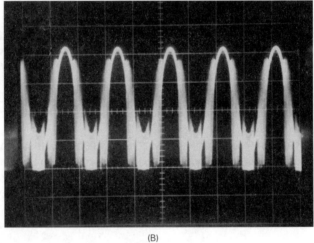

(B)

FIGURE 1-9 (A) Using series diodes to protect the op-amp against reverse polarity conditions; (B) output signal when power supply lines are not decoupled; (C) same signal with decoupling capacitors present; (D) expansion of oscilloscope trace to show the oscillation.

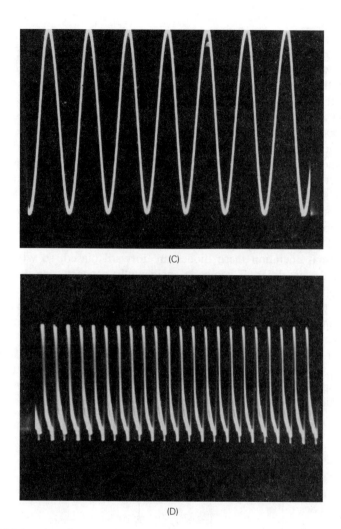

(C)

(D)

FIGURE 1-9 (continued)

performance of these capacitors drops drastically as frequency increases, and may be very poor at higher frequencies that are nonetheless within the range of the device. At those high frequencies the typical electrolytic capacitor is ineffective. For this reason we also sometimes use a smaller value capacitor, but one that is of a type that will work at higher frequencies (e.g. mylar, mica, ceramic, etc.). This situation is changing, however, because certain new forms of capacitor are now available that offer high frequency operation as well as high capacitance.

Figures 1-9B and 1-9C show the output waveform of a high frequency operational amplifier in both non-decoupled (Fig. 1-9B) and decoupled

(Fig. 1-9C) configurations. In Fig. 1-9B we see a 400 Hz sinusoidal waveform that has a high frequency oscillation superimposed on it. In Fig. 1-9C, however, we see the same waveform after the decoupling capacitors were added to the $V-$ and $V+$ pins of the operational amplifier used in this experiment. Note that the oscillation is removed. In Fig. 1-9D we see the actual oscillation (with 400 Hz sinewave turned off) in an expanded oscilloscope timebase range.

One practical 'rule of thumb' ensures the success of the circuit with regards to noise and oscillation: *place those capacitors as close as physically possible to the body of the amplifier*. The 0.1 μF capacitors (*C*1 and *C*2) are more important than the higher value capacitors, so should be closest to the IC amplifier body.

Protection against reverse polarity conditions is shown in Fig. 1-9 also. Diodes *D*1 and *D*2 are placed in series with each DC power supply line. Under normal operation these diodes are forward biased, so will conduct current to the amplifier. If someone accidentally connects one or both DC power supplies backwards, then these diodes are reverse biased so will not conduct current. Thus, the series diodes protect the amplifier IC from incorrect power supply connection. Typical diodes for this application are any of the 1N400*x* series (1N4001 through 1N4007).

Another method for protecting the device is shown in Fig. 1-10. In this case, a zener diode is placed across the two DC power supply terminals of the amplifier. The zener potential must be greater than the maximum actual value, but less than the maximum permissible value, of the quantity

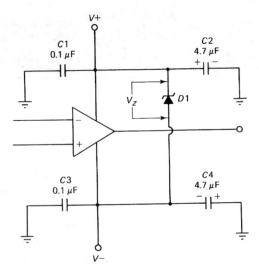

FIGURE 1-10   Use of a zener diode between *V*– and *V*+ to protect against polarity reversal.

$[(V+) - (V-)]$. For the case where DC power supplies of ± 12 volts are used, then this value is 24 volts. A 28-volt zener diode would be adequate if the maximum permissible value is specified as 30 volts (provided that the power supply voltages are reasonably stable). Under these conditions, with V_z greater than the voltage between the terminals, zener diode $D1$ is reverse biased at a voltage lower than the zener potential. Thus, it is not conducting in normal operation. In reverse polarity operation, diode $D1$ becomes forward biased in the normal non-zener mode. It will pass current around the amplifier harmlessly. If excessive $(V-)$-to-$(V+)$ voltage is applied, then the zener diode will conduct and clamp the voltage to V_z.

The protection of the multiple stage amplifier is shown in Fig. 1-11. There are two alternatives shown in this figure. In one case, 1N400x-series diodes in reverse bias state are placed across the DC power supply lines and current-limiting resistors in series to prevent them from burning up. The diodes are normally reverse biased, but when one or both DC power supplies are reversed, then these diodes become forward biased and short the line to ground. The second alternative is to place the diodes in series with the line at the power supply terminals of the IC device (shown in dotted lines in Fig. 1-11). This method is analogous to the method of Fig. 1-9, except that it serves more than one amplifier.

1-8.2 DC power supplies for laboratory experiments

Many readers will perform laboratory experiments independently, as part of their school course work, or they will construct electronic projects. Guidelines for suitable DC power supplies are given below.

The power supply selected should offer either a single bipolar power supply or two independent 12 volt DC supplies that are not ground referenced. The 'non-grounded' feature allows you to create a bipolar supply by connecting the positive output terminal of one supply to the negative output terminal of the other (see Fig. 1-12).

Desirable features to have on a bench power supply include the following:

1. output voltages fixed at ± 12 volts DC (or ± 15 volts); or
2. output voltages adjustable from zero to greater than 12 volts DC;
3. current available from each polarity not less than 100 milliamperes;
4. metered outputs (I and V);
5. voltage regulation;
6. current-limiting for output short circuit protection; and
7. overvoltage protection.

Although not all good selections will have all of these features, those that do are clearly superior for most laboratory applications.

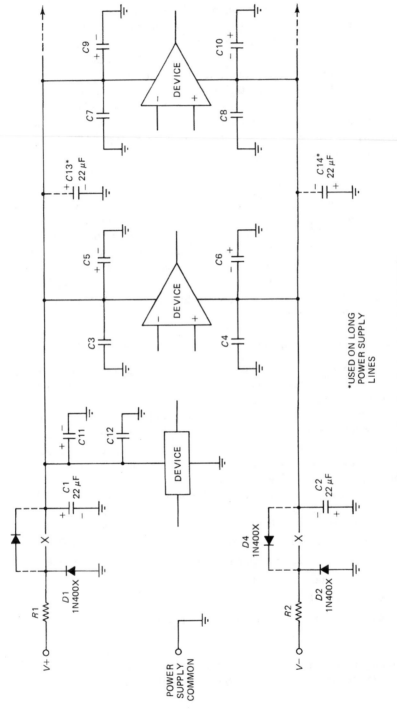

FIGURE 1-11 Using rectifier diodes to protect a series of semiconductor devices in a circuit.

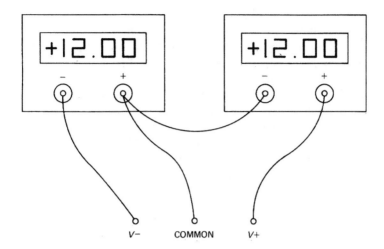

V−   COMMON   V+

FIGURE 1-12 Using standard bench DC power supplies for linear IC and op-amp applications.

1-9 LABORATORY CONSTRUCTION PRACTICES ───────────

Successfully performing student experiments, or designing and de-bugging circuits on a professional level, requires certain skills and knowledge of construction practices. In this section we will take a look at some of these practical matters before proceeding with the rest of the text.

There are various levels of preproduction construction practice, of which we can identify four major categories: (1) *partial breadboard*, (2) *complete breadboard*, (3) *brassboard*, and (4) *full-scale development model*.

The two 'breadboard' categories involve using special construction methods to check circuit validity and make certain preliminary measurements. The breadboard is in no way usable in actual equipment, but rather is a bench or laboratory device.

The difference between partial and complete breadboards is merely one of scale. The partial breadboard is used to check out a partial circuit, or a small group of circuits. The complete breadboard will be the full-up circuit, and is much more extensive than the partial breadboard. There are fundamental differences in the type of breadboarding hardware used for both types, especially in large circuits.

The two most commonly used types of prototyping or breadboarding hardware are the push-pin breadboard and wirewrapping board. A push-pin breadboard consists of numerous multi-pin socket strips in which component leads and #22 solid hook-up wire can be inserted. Some are bare sockets, while others include sockets but also offer built-in DC power supplies: typically, +12 volts at 100 mA, −12 volts at 100 mA and +5 volts at 1.5 A (for TTL digital logic IC devices). Located between the large multi-pin IC sockets are bus sockets in which all pins in a given row are strapped together. These sockets are generally used for power distribution and common ('ground') busses. Interconnections between sockets, power sources, and components are made using #22 insulated hook-up wire. The bared ends of the wires are pressed into the socket holes.

The other method of breadboarding is wirewrapping. Sockets with special square or rectangular wirewrap pins are mounted on a piece of 'universal' printed circuit board. These boards are perforated to accept wirewrap IC sockets, and usually have 'printed' power distribution and ground busses. Using a special wirewrapping tool, the wire interconnects are strapped from point-to-point as needed by the circuit. The wire is wrapped around each contact tightly enough to break through the insulation to make the electrical connection.

In general, socketed breadboards are typically used for student applications, for partial breadboarding in professional laboratories, and for small design projects that don't justify the cost of the wirewrap board. Wirewrapping is used for larger, more extensive, projects. Although the distinctions between types of breadboard, and the respective supporting hardware, are quite flexible, they hold true for a wide range of situations.

Brassboards are full-up model shop versions of the circuit that will plug into the actual equipment cabinet in which the final circuit will reside. Some brassboards are wirewrapped, but most are printed circuits. The important consideration is the that the brassboard is near its final configuration and can be tested *in situ* in the actual equipment. In contrast, the breadboard is purely a laboratory model. The brassboard is usually made using model shop handwork methods, and not routine automated production methods. The brassboard differs from the final version in that it might contain 'dead bug' and 'kluge card' modifications in which components are informally mounted on the board for purposes of testing circuit changes.

The full-scale development (FSD) model is the highest preproduction model. It might look very much like the first article production model that is eventually made, and is intended for field tests of the final product. For example, an FSD model of a two-way landmobile radio transceiver may be mounted in an actual vehicle and used for its intended purpose in tests or field trials. For aircraft electronics, the FSD model must be built and tested according to flight worthiness criteria. The FSD model is built as near to regular production methods as possible.

There are several principles to remember when breadboarding electronic circuits. While it is possible to get away with ignoring these rules, they constitute good practice and ignoring them is risky. The rules are:

1. Insert and remove components only with the power turned off.
2. Wiring and changes to wiring are done with the power turned off.
3. Check all wiring prior to applying power for the first time.

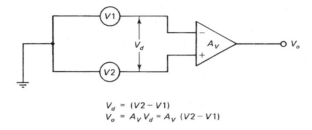

$$V_d = (V2 - V1)$$
$$V_o = A_V V_d = A_V (V2 - V1)$$

FIGURE 1-13 Differential signals represented as the difference between two single-ended signals.

4. Power distribution and ground wiring is always done first, and then checked before any other wiring is done (note: in many schematics the $V-$ and $V+$ wiring is omitted, but that doesn't mean these connections are not to be made).

5. Use single point (or 'star') grounding wherever possible. A ground plane or ground bus should not be used unless signal frequencies are low, signal voltage levels are relatively large, and current drain from the DC power supply is low. Otherwise, ground noise and 'ground loop' voltage drop problems will exist.

6. Applying a signal to IC input pins when DC is not applied can cause the substrate 'diode' to be forward biased, with the potential for damage to the device being very high. Therefore: (a) do not apply signal until the DC power is turned on, (b) turn off the signal source prior to turning off the DC power supply, and (c) never apply a signal with a positive peak that exceeds $V+$ or a negative peak that exceeds $V-$.

7. All measurements are to be made with respect to common or ground, unless special test equipment is provided. For example, in Fig. 1-13 differential voltage V_d is composed of two ground referenced voltages, $V1$ and $V2$. Measure these voltages separately and then take the difference between them (i.e. $V2 - V1$).

In this chapter we introduced the integrated circuit, discussed the power supply for the IC, and demonstrated some construction practices and techniques. In the next chapter we will introduce the ideal and practical operational amplifier, and examine the applications of an op-amp with no feedback network.

1-10 SUMMARY

1. 'Universal' IC amplifiers include devices such as the operational amplifier, current difference or 'Norton' amplifier, and the operational transconductance amplifier.

2. Analog signals are continuous in both range and domain; digital signals are discrete in range and possibly also domain; sampled signals are continuous in range but discrete in domain.

3. Common forms of integrated circuit package include the metal can, dual in-line package (DIP), flat-pack, and surface mounted device.

4. There are several 'scales of integration' available: Small Scale Integration (<20 internal devices); Medium Scale Integration (approximately 100 internal devices); Large Scale Integration (100 to 1000 internal devices); and Very Large Scale Integration (more than 1000 internal devices).

5. Hybrid devices are a cross between integrated and discrete technology. Treated as if they are an IC device, these hybrids are actually packaged miniature printed circuits.

6. Most linear IC devices operate from bipolar power supplies; i.e. a supply in which $V-$ is negative to common, while $V+$ is positive to common. By special biasing methods, the linear device can operate from single polarity supplies.

7. Both 'socketed breadboards' and wirewrapping technique are used to prototype circuits.

1-11 RECAPITULATION

Now return to the objectives and Pre-quiz questions at the beginning of the chapter and see how well you can answer them. If you cannot answer certain questions, place a check mark to each and review the appropriate parts of the text. Next, try to answer the questions and work the problems below, using the same procedure.

1-12 QUESTIONS AND PROBLEMS

1. In an _____ amplifier device the voltage amplifier transfer function is controlled entirely by the feedback loop characteristics.

2. List two principal advantages of integrated versus discrete electronics components.

3. A _____ amplifier has two inputs: inverting and noninverting. This type of amplifier forms the basis for the popular operational amplifier.

4. A Norton amplifier is also called a _____ _____ amplifier.

5. The _____ _____ amplifier has a transfer function of the basic form $\Delta I_o / V_{in}$.

6. A signal is found to be continuous in both range and domain. This is a(n) _____ signal.

7. A signal is found to be discrete in both range and domain. This is a(n) _____ signal.

8. A signal is found to be continuous in range, but discrete in domain. This is a(n) _____ _____ signal.

9. List three types of package used for IC linear devices.

10. True or false: all circuits involving linear IC devices have an output signal that is related to the input signal by a linear scaler factor.

11. A device contains 12 internal components. This device is _____ scale integration.

12. A device contains 100 internal components. This device is _____ scale integration.

13. A device contains 684 internal components. This device is _____ scale integration.

14. The substrate diode must be kept normally _____ biased.

15. An operational amplifier has specified voltages as follows: maximum $V- =$ -18 volts, maximum $V+ = +18$ volts, and maximum pin-to-pin $[(V-)$ to $(V+)]$ voltage of 28 volts DC. If $V+ = 15$ volts, what is the maximum permissible value of $V-$?

16. What is the maximum permissible input voltage peak value when the maximum applied $V+ = 18$ volts?

17. What is the value of voltage $V1$ in Fig. 1-14?

18. Using the standard engineering practice 'rule of thumb', what is the minimum allowable value of $C1$ in Fig. 1-14 if the amplifier must operate down to a frequency of 30 Hz?

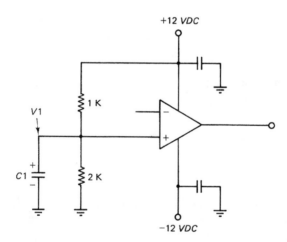

FIGURE 1-14

CHAPTER 2

The IC operational amplifier

OBJECTIVES

1. Learn the basic pin functions of the standard operational amplifier.
2. Learn the most common package styles used for IC operational amplifiers.
3. Describe the electrical parameters associated with each basic pin on an operational amplifier.
4. Learn the properties of the ideal operational amplifier.

2-1 PRE-QUIZ

These questions test your prior knowledge of the material in this chapter. Try answering them before you read the chapter. Look for the answers (especially those you answered incorrectly) as you read the text. After you have finished studying the chapter try answering these questions again, and those at the end of the chapter (see Section 2-11).

1. List the seven properties of the 'ideal' operational amplifier.
2. List the five basic pins found on most operational amplifiers.
3. Describe the difference between inverting and noninverting inputs.
4. Write the transfer function for a differential amplifier in which A_v is the gain, and $V1$ and $V2$ are the input signals applied to the inverting and noninverting inputs, respectively.

2-2 INTRODUCTION

The original operational amplifiers (1948) were designed to perform *mathematical operations* in analog computers. Their use in other applications follows from the fact that the op-amp is basically a very good DC differential amplifier with an extremely high gain. The immense benefits and extreme flexibility of the device derive from that one simple fact. Applications for the operational amplifier are found in instrumentation, process monitoring and control, servo control systems, signals processing, communications, measuring and testing circuits, alarm systems, medicine, science and even in some digital computers. It is only somewhat exaggerated to call the operational amplifier a universal linear amplifier.

2-3 OPERATIONAL AMPLIFIERS

Operational amplifiers are probably the most widely used linear integrated circuits on the market today. They are, without a doubt, the most flexible linear IC devices available because we can manipulate the overall forward transfer function by manipulating the feedback network properties. Other 'universal' amplifiers (e.g. CDA and OTA) have failed to overcome the op-amp in either sales or usefulness.

Figure 2-1 shows a simplified schematic diagram of the internal circuitry of the popular, and low-cost, 741 operational amplifier device. There are three main sections to this circuit: *input amplifier*, *gain stages*, and *output amplifier*.

The output stage is a complementary symmetry push-pull DC power amplifier that produces from 50 to 500 milliwatts of output power depending on the design. The output stage operates as a push-pull amplifier because transistors $Q9$ and $Q10$ have opposite polarities: $Q9$ is NPN and $Q10$ is PNP. Output signal is taken from the junction of the $Q9/Q10$ emitters. If the two transistors conduct equal amounts, then the net output voltage is zero.

The intermediate gain stages are shown here in simplified block diagram form. These stages provide the high gain required, some level translation, and (in the case of the 741) some internal frequency compensation.

The input stage is a DC differential amplifier made from bipolar transistors. Although there are a few single-input devices on the market sold as 'operational' amplifiers, they are in reality high gain DC amplifiers, not true 'op-amps'. The reason is that a true *operational* amplifier must be able to perform a wide range of mathematical operations, and that ability requires both inverting and noninverting input functions. This same reason is also used to explain why bipolar DC power supplies must be used in op-amp circuits: results (i.e. output voltages) may be either zero, positive or negative... and the device must be able to accommodate all three of these possibilities.

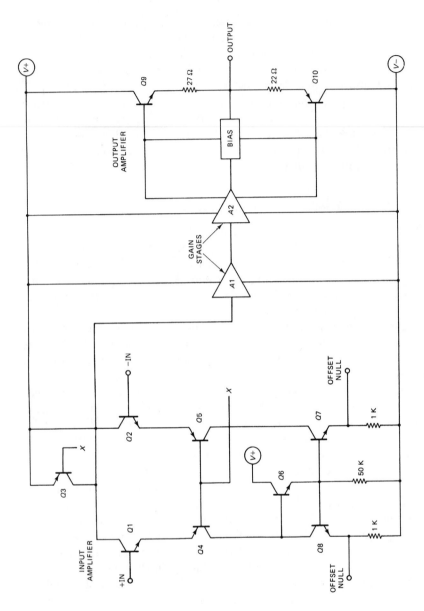

FIGURE 2-1 Typical internal circuitry for operational amplifier.

2-3.1 DC differential amplifiers

Figure 2-2A shows a simplified DC differential amplifier circuit. Two transistors ($Q1$ and $Q2$) are connected at their emitters to a single *constant current source* (CCS), $I3$. Because current $I3$ cannot vary, changes in either $I1$ or $I2$ will also affect the other current; it's a zero-sum situation. For example, an increase in current $I1$ means a necessary decrease in current $I2$ in order to satisfy Eq. (2-1):

$$I3 = I1 + I2 \qquad (2\text{-}1)$$

where:

$I3$ is the constant current
$I1$ is the C–E current of transistor $Q1$
$I2$ is the C–E current of transistor $Q2$

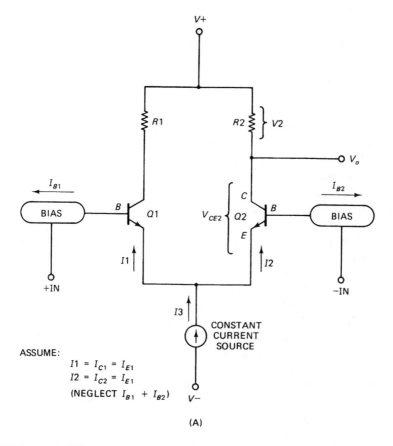

(A)

FIGURE 2-2   (A) Differential input stage; (B) equivalent circuit.

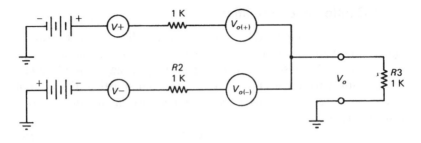

ASSUME:

$$V+ = +10 \, VDC$$
$$V- = -10 \, VDC$$
$$V_o = V_{o(+)} + V_{o(-)}$$
$$R1 = R2 = R3 = 1 \, K\Omega$$

A) $V_{o(-)} = \dfrac{(V-)(R3)}{(R2+R3)} = \dfrac{(-10 \, Vdc)(1 \, K)}{(1 \, K + 1 \, K)} = \dfrac{(-10 \, Vdc)(1)}{2} = -5 \, Vdc$

B) $V_{o(+)} = \dfrac{(V+)(R3)}{(R1+R3)} = \dfrac{(+10 \, Vdc)(1 \, K)}{(1 \, K + 1 \, K)} = \dfrac{(+10 \, Vdc)(1)}{2} = +5 \, Vdc$

C) $V_o = V_{o(-)} + V_{o(+)} = (-5 \, Vdc) + (+5 \, Vdc) = 0 \, Vdc$

(B)

FIGURE 2-2 (continued)

Although the collector and emitter currents for each transistor are not actually equal to each other by the amount of the base current, we may neglect I_{B1} and I_{B2} for the time being. For purposes of this discussion we may assume that $I1 = I_{C1} = I_{E1}$, and $I2 = I_{C2} = I_{E2}$ (even though they are not, in fact, equal in real circuits).

Consider two voltage drops created by current $I2$ (which is the C–E current flowing in transistor $Q2$): $V2$ is the voltage drop $I2 \times R2$, while V_{CE2} is the collector–emitter voltage drop across $Q2$. Voltage V_{CE2} depends on the conduction of $Q2$, which in turn is determined by the signal applied to the noninverting input ($+IN$).

If the signal voltages applied to $-IN$ and $+IN$ inputs are equal, then $I1$ and $I2$ are equal. In that case, the quiescent values of $V2$ and V_{CE2} are approximately equal (as determined by internal bias networks) so the relative contributions of $V-$ and $V+$ to output potential V_o are equal. How does this work? A model circuit is shown in Fig. 2-2B to demonstrate how V_o is formed. The output voltage V_o is the sum of two contributors: $V_{o(-)}$ is the contribution from $V-$, and $V_{o(+)}$ is the contribution from $V+$. These voltages are derived from the voltage drops across $R1$ and $R2$, respectively, and in our model represent the voltage drops $V2$ and V_{CE2} above. As long as $R1$ and $R2$ are balanced, then the sum $V_o = V_{o(-)} + V_{o(+)}$ is equal to zero. But if either $R1$ or $R2$ change, then V_o will be non-zero. If you wish to prove this fact to yourself, then repeat the arithmetic shown in Fig. 2-2B using a different value than 1 kohm for either $R1$ or $R2$. For example, when $R1$ is changed to 2 kohms, then the output voltage V_o is -1.667 volts DC instead of zero.

Now let's consider the operation of the inverting input ($-$IN), which is the base terminal of transistor $Q2$. Recall that an inverting input produces an output signal that is 180 degrees out of phase with the input signal. In other words, as V_{-IN} goes positive, V_o goes negative; as V_{-IN} goes negative, V_o goes positive.

If a positive signal voltage is applied to the $-$IN input, and $+$IN $= 0$, then NPN transistor $Q2$ is turned on harder. The effect is an increase in current $I2$ because the collector–emitter resistance of $Q2$ drops. We now have an inequality between $V2$ and V_{CE2}: voltage $V2$ increases, while V_{CE2} decreases. The result is that the contribution of $V-$ to V_o is greater, so V_o goes negative. The base of $Q2$ is, therefore, the inverting input because *a positive input voltage produced a negative output voltage.*

If the signal voltage applied to the $-$IN input is negative instead of positive, then the situation changes. In that case, $Q2$ starts to turn off, so $I2$ drops. Voltage V_{CE2} therefore increases and $V2$ decreases. The relative contribution of $V+$ to V_o is now greater than the contribution of $V-$, so V_o goes positive. Again inverting behavior is seen: *a negative input voltage produced a positive output voltage.*

Now consider the noninverting input, which is the base of transistor $Q1$. Recall that a noninverting input produces an output signal that is in-phase with the input signal. A positive going input signal produces a positive going output signal, and a negative going input signal produces a negative going output.

Suppose $-$IN $= 0$, and a positive signal voltage is applied to $+$IN. In this case $I1$ increases. Because Eq. (2-1) must be satisfied, an increase in $I1$ results in a decrease of $I2$ in order to keep $I3 = I1 + I2$ constant. Reducing $I2$ reduces $V2$ (which is $I2 \times R2$), so the contribution of $V+$ to V_o goes up: V_o *goes positive in response to a positive input voltage.* This is the behavior expected of a noninverting input.

Now suppose that a negative signal voltage is applied to $+$IN. Now there is a decrease in the conduction of $Q1$, so $I1$ drops. Again, to satisfy Eq. (2-1) current $I2$ increases. This condition increases $V2$, reducing the contribution of $V+$ to V_o, forcing V_o negative. Because a *negative input voltage produced a negative output voltage,* we may again affirm that $+$IN is a noninverting input.

2-3.2 Categories of operational amplifiers

Now that we have discussed the basic operational amplifier let's widen our discussion a bit to encompass a larger selection of devices. Table 2-1 shows a hierarchy of commonly available devices. Some of these devices have been

TABLE 2-1 Classification of some common operational amplifiers.

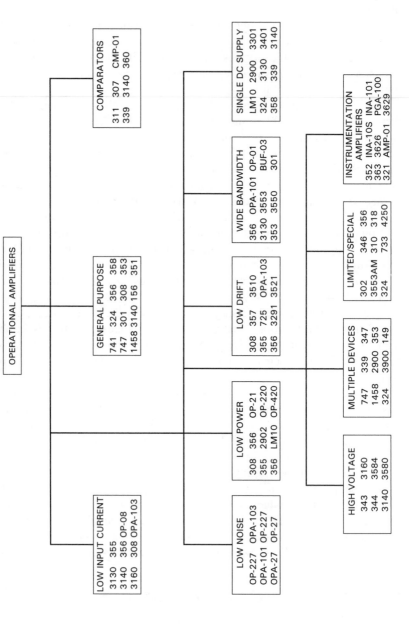

on the market for a long time, while others are relatively new. This list is intended to be representative, rather than exhaustive. Let's take a brief look at each category.

General purpose. These are 'garden variety' operational amplifiers that are neither special purpose nor premium devices. Most of these devices are 'frequency compensated', so designers trade off bandwidth for inherent stability. As such, the general purpose devices can be used in a very wide range of applications with very few external components. Devices are usually selected from this category unless a property of another class brings unique advantage to some particular application.

Voltage comparators. These devices are not strictly speaking 'operational amplifiers', but are based on op-amp circuitry. While all op-amps can be used as voltage comparators, the reverse is not true: IC comparators (e.g. LM-311) cannot usually be used as op-amps. More about comparator circuits will be said in Section 2-7.

Low input current. Although ideal op-amps (Section 2-4) have zero input bias current, real devices have a small current due to input transistor biasing or leakage. This class of devices typically uses MOSFET, JFET or superbeta (bipolar Darlington) configuration transistors for the input stage instead of single NPN/PNP bipolar devices. The manufacturer may choose to use a nulling or selection technique in-process that reduces input bias current. Low input current devices typically have picoampere level currents, rather than microampere or milliampere input currents found in some other devices.

Low noise. These devices are optimized to reduce internally generated noise.

Low power. This category of op-amp optimizes internal circuitry to reduce power consumption. Many of these devices also operate at very low DC power supply potentials (e.g. ±1.5 volts DC).

Low drift. All DC amplifier circuits experience an erroneous change of output voltage as a function of temperature. Devices in this category are internally compensated to minimize temperature drift. These devices are typically used in instrumentation circuits where drift is an important concern, especially when handling low level input signals where the magnitude of the drift component could be of the same order as the signal.

Wide bandwidth. Also called 'video' op-amps in some literature, these devices have a very high gain–bandwidth (G–B) product. One Burr-Brown device, for example, has a G–B product of 600 MHz, compared with 0.300 to 1.2 MHz for various 741-family devices.

Single DC supply. These devices are able to provide op-amp-like behavior from a monopolar (typically $V+$) DC power supply. However, not all op-amp performance will be available from some of these devices because the output voltage may not be able to assume negative values.

High voltage. Most op-amps operate from ±6 to ±22 volts DC power supplies. A few devices in the high voltage category operate from ±44 volts DC supplies, and at least one proprietary hybrid model (not listed here) operates from ±100 volts DC power supplies.

Multiple devices. This category requires only that more than one op-amp be included in the same package. Devices exist with two, three or four operational amplifiers in a single package. The 1458 device, for example, contains two 741-family devices in either 8-pin metal can or miniDIP packages.

Limited/special purpose. The devices in this category are designed for either a limited range of uses, or highly specialized uses. The type 302 buffer, for example, is an op-amp that is connected internally into the noninverting unity gain follower configuration (see Chapter 3). Some consumer audio devices also fit into this category.

Instrumentation amplifiers. Although the instrumentation amplifier (IA) is arguably a 'special purpose' device, it is sufficiently universal to warrant a class of its own. The IA is a DC differential amplifier made of either two or three internal op-amps. Voltage gain can be set by either one or two external resistors. Chapter 6 deals with these amplifiers in detail.

2-4 THE `IDEAL` OPERATIONAL AMPLIFIER ——————

When you study any type of electron device, be it vacuum tube, transistor or integrated circuit, it is wise to start with an ideal representation of that device and then proceed to practical devices. In some cases, the practical and ideal devices are so far apart that you might wonder at the wisdom of this approach. In engineering and technical schools the students often do laboratory experiments to put into practice what was learned in the theory classes. When students study their lab workbook sheets they sometimes wonder why the experiments fail to come close to predicted theoretical behavior. But IC operational amplifiers, even low-cost commercial products so nearly approximate the ideal op-amp of textbooks that the lab experiments actually work. The 'ideal model' analysis method thus becomes extremely useful for understanding the technology, learning to design new circuits, or figuring out how someone else's circuit works.

In Chapter 3 we will discuss the inverting and noninverting amplifier configurations of the op-amp. In those discussions we will derive the design equations that describe the operation of real circuits from both the ideal model and a feedback amplifier model. The usefulness of our simplified approach proceeds directly from the correspondence of the ideal and practical operational amplifier IC devices.

2-4.1 Properties of the ideal operational amplifier

The ideal op-amp is characterized by seven properties. From this short list of properties we can deduce circuit operation and design equations. Also, the list gives us a basis for examining non-ideal operational amplifiers and their defects (plus solutions to the problems caused by those defects). The basic properties of the op-amp are:

1. infinite open-loop voltage gain;
2. infinite input impedance;
3. zero output impedance;
4. zero noise contribution;
5. zero DC output offset;
6. infinite bandwidth; and
7. both differential inputs stick together.

Let's take a look at these properties to determine what they mean in practical terms. You will find that some real op-amps only approximate some of these ideals, but for others on the list the approximation is extremely good.

Property No. 1 — infinite open-loop gain. The open-loop gain of any amplifier is its DC gain without either negative or positive feedback. By definition, negative feedback is a signal fed back to the input 180° out of phase. In operational amplifier terms this means feedback between the output and the inverting input.

Negative feedback has the effect of reducing the open loop gain (A_{vol}) by a factor (called 'β') that depends upon the transfer function and properties of the feedback network. Figure 2-3 shows the basic configuration for any negative feedback amplifier. The 'transfer equation' for any circuit is the output function divided by the input function. The transfer function of a voltage amplifier is, therefore, $A_{vol} = V_o/V_{in}$. In Fig. 2-3 the term A_{vol}

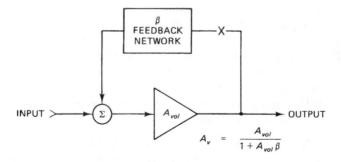

FIGURE 2-3   Block diagram of a feedback amplifier.

represents the gain of the amplifier element only, i.e. the gain with the feedback network disconnected. The term β represents the transfer function of the feedback network without the amplifier. The overall transfer function of this circuit, i.e. with both amplifier element and feedback resistor in the loop, is defined as:

$$A_v = \frac{A_{vol}}{1 + A_{vol}\beta} \qquad (2\text{-}2)$$

where:

A_v is the closed-loop (feedback network connected) gain

A_{vol} is the open-loop (feedback network disconnected) gain

β is the transfer function for the feedback network

In the ideal op-amp A_{vol} is infinite, so the voltage gain is a function only of the feedback network. In real op-amps, the value of the open-loop gain is not infinite, but it is quite high. Typical values range from 20 000 in low-grade consumer audio models to more than 2 000 000 in premium units (typically 200 000 to 300 000).

Property No. 2 — infinite input impedance. This property implies that the op-amp input will not load the signal source. The input impedance of any amplifier is definable as the ratio of the input voltage the input current: $Z_{in} = V_{in}/I_{in}$. When the input impedance is infinite, therefore, we must assume that the input current is zero. Thus, an important implication of this property is that the operational amplifier inputs neither sink nor source current. In other words, it will neither supply current to an external circuit, nor accept current from an external circuit. We will depend upon an implication of this property ($I_{in} = 0$) to perform the circuit analysis in Chapter 3.

Real operational amplifiers have some finite input current other than zero. In low-grade devices this current can be substantial (e.g. 1 mA in the 741-family), and will cause a large output offset voltage error in medium and high gain circuits. The primary source of this current is the base bias currents from the NPN and PNP bipolar transistors used in the input circuits. Certain premium grade op-amps that feature bipolar inputs reduce this current to nanoamperes or picoamperes. In op-amps that use field effect transistors (FET) in the input circuits, on the other hand, the input impedance is quite high due to the very low leakage currents normally found in FET devices. The JFET input devices are typically called 'BiFET' op-amps, while the MOSFET input models are called 'BiMOS' devices. The CA-3140 device is a BiMOS op-amp in which the input impedance approaches 1.5 terraohms (i.e. 1.5×10^{12} ohms) — which is near enough to 'infinite' to make the inputs of those devices approach the ideal.

Property No. 3 — zero output impedance. A voltage amplifier (of which class, the op-amp is a member) ideally has a zero output impedance. All real voltage amplifiers, however, have a non-zero output impedance. Figure 2-4

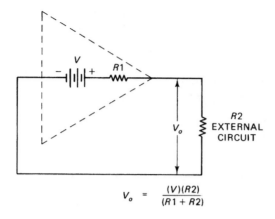

$$V_o = \frac{(V)(R2)}{(R1 + R2)}$$

FIGURE 2-4   Equivalent circuit of voltage amplifier output.

represents any voltage source (including amplifier outputs) and its load (external circuit). Potential V is a perfect internal voltage source with no internal resistance; resistor $R1$ represents the internal resistance of the source, and $R2$ is the load. Because the internal resistance (which in amplifiers is usually called 'output resistance') is in series with the load resistance, the output voltage V_o that is available to the load is reduced by the voltage drop across $R1$. Thus, the output voltage is given by:

$$V_o = \frac{V\,R2}{R1 + R2} \tag{2-3}$$

It is clear from the above equation that the output voltage will equal the internal source voltage only when the output resistance of the amplifier $(R1)$ is zero. In that case, $V_o = V \times (R2/R2) = V$. Thus, in the ideal voltage source, we get the maximum output voltage (and the least error) because no voltage is dropped across the internal resistance of the amplifier.

Real operational amplifiers do not have a zero output impedance. The actual value is typically less than 100 ohms, with many being in the neighborhood of 30 to 50 ohms. Thus, for typical devices the operational amplifier output can be treated as if it were ideal.

A rule of thumb used by designers is to set the input resistance of any circuit that is driven by a non-ideal voltage source output to be at least ten times the previous stage output impedance (the higher the better). We see this situation in Fig. 2-5. Amplifier $A1$ is a voltage source that drives the input of amplifier $A2$. Resistor $R1$ represents the output resistance of $A1$ and $R2$ represents the input resistance of amplifier $A2$. In practical terms, we find that the circuit where $R2 > 10R1$ will yield results acceptably close to 'ideal' for many purposes. In some cases, however, the rule $R2 > 100R1$ or $R2 > 1000R1$, must be followed where greater gain accuracy is required.

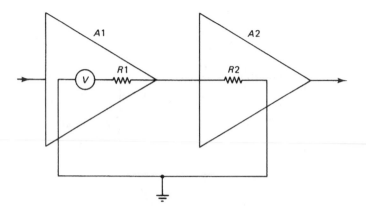

FIGURE 2-5   Equivalent circuit of cascaded voltage amplifiers.

Property No. 4 — zero noise contribution. All electronic circuits, even simple resistor networks, produce noise signals. A resistor creates noise due to the movement of electrons in its internal resistance element material. For all resistive circuits, the noise contribution is:

$$V_{\text{noise}} = \sqrt{4KTBR} \tag{2-4A}$$

$$P_{\text{noise}} = 4KTBR \tag{2-4B}$$

where:

V_{noise} is the noise potential in volts (V)

P_{noise} is the noise power in watts (W)

K is Boltzman's constant (1.38×10^{-23} J/K)

T is the temperature in Kelvin (K)

B is the bandwidth in hertz (Hz)

R is the resistance in ohms (Ω)

As you can see from Eq. (2-4), the noise voltage produced by a resistor varies with the square root of its resistance, while noise power varies directly with resistance.

In the ideal operational amplifier, zero noise voltage is produced internally. Thus, any noise in the output signal must have been present in the input signal as well. Except for amplification, the output noise voltage will be exactly the same as the input noise voltage. In other words, the op-amp contributed nothing extra to the output noise. This is one area where practical devices depart quite a bit from the ideal. Practical op-amps do not approximate the ideal, except for certain high-cost premium low-noise models.

Amplifiers use semiconductor devices that create not merely resistive noise (as described above), but also create special noise of their own. There are a number of internal noise sources in semiconductor devices, and any good text on transistor theory will give you more information on them. For present purposes, however, assume that the noise contribution of the op-amp can be considerable in low signal level situations. Premium op-amps are available in which the noise contribution is very low, and these devices are usually advertised as premium low-noise types. Others, such as the metal can version of the CA-3140 device, will offer relatively low-noise performance when the DC supply voltages are limited to ±5 Vdc, and the metal package of the op-amp is fitted with a flexible 'TO-5' style heatsink.

Property No. 5 — zero output offset. The output offset voltage of any amplifier is the output voltage that exists when it should be zero. The voltage amplifier sees a zero input voltage when the inputs are both grounded. This connection should produce a zero output voltage. If the output voltage is non-zero, then there is an output offset voltage present. In the ideal op-amp, this offset voltage is zero volts — but real op-amps exhibit at least some amount of output offset voltage. In the real IC operational amplifier the output offset voltage is non-zero, although it can be quite low in some cases. There are several methods for dealing with output offset voltage, and these will be discussed at length in Chapter 5.

Property No. 6 — infinite bandwidth. The 'ideal' op-amp will amplify all signals from DC to the highest AC frequencies. In real op-amps, however, the bandwidth is sharply limited. There is a specification called the 'gain–bandwidth (G–B) product', which is symbolized by F_t. This specification is the frequency at which the voltage gain drops to unity (1). The maximum available gain at any frequency is found from dividing the maximum required frequency into the gain–bandwidth (G–B) product. If the value of F_t is not sufficiently high, then the circuit will not behave in classical op-amp fashion at some frequencies.

Some common op-amps have G–B products in the 10 to 20 MHz range, while special-purpose devices have G–B products of 500 to 1000 MHz. Other devices, on the other hand, are quite frequency limited. The 741-family of devices is very limited, such that the device will perform as an op-amp only to frequencies of a few kilohertz. Above that range, the gain drops off considerably. But in return for this apparent limitation, we obtain unconditional stability; such op-amps are said to be 'frequency compensated'. It is the frequency compensation of these devices that both reduces the G–B product and provides the inherent stability. Non-compensated op-amps will yield wider frequency response, but only at the expense of a tendency to oscillate. Those op-amps may spontaneously oscillate without any special encouragement unless good circuit design and construction practices are followed.

Property No. 7 — differential inputs stick together. Most operational amplifiers have two inputs: one inverting (−IN) input and one noninverting (+IN) input. 'Sticking together' means that a voltage applied to one of these inputs also appears at the other input. This voltage is real; it is not merely some theoretical device used to evaluate circuits. If you apply a voltage to, say, the inverting input, and then connect a voltmeter between the noninverting input and the power supply common, then the voltmeter will read the same potential on the noninverting as it did on the inverting input. The implication of this property is that we must treat both inputs the same mathematically. This fact will make itself felt when we discuss the concept of 'virtual' as opposed to actual grounds, and again when we deal with the noninverting follower circuit configuration.

Our discussion thus far has not dealt with the problems caused by the departures from ideal properties found in real operational amplifiers. For that discussion you may turn to Chapter 5. In the meantime, however, we will want to first develop the two major configurations of the operational amplifier: *inverting follower* and *noninverting follower* (Chapter 3). The inverting follower produces an output signal that is 180° out of phase with its input signal. The noninverting follower, as you might expect, produces an output signal that is in-phase with its input signal. Both of these circuits are covered in Chapter 3. Almost all other operational amplifier circuits are variations on either inverting or noninverting follower circuits. Understanding these two configurations will allow you to understand, and either design or modify, a wide variety of different circuits using IC operational amplifiers.

2-5 STANDARD OPERATIONAL AMPLIFIER PARAMETERS ———

Understanding operational amplifier circuits requires knowledge of the various parameters given in the specification sheets. The list below represents the most commonly needed parameters. Methods of measuring some of these parameters are discussed in Chapter 5.

Open-loop voltage gain (A_{vol}). Voltage gain is defined as the ratio of output voltage to input signal voltages (V_o/V_{in}), which is a dimensionless quantity. The open-loop voltage gain is the gain of the circuit without feedback (i.e. with the feedback loop open). In an ideal operational amplifier A_{vol} is infinite, but in practical devices it will range from about 20 000 for low-cost devices to over 2 000 000 in premium devices; 741-family devices offer gains of about 300 000.

Large signal voltage gain. This gain figure is defined as the ratio of the maximum allowable output voltage swing (usually one to several volts less than $V-$ and $V+$) to the input signal required to produce a swing of ±10 volts (or some other standard).

Slew rate. This parameter specifies the ability of the amplifier to transition from one output voltage extreme to the other extreme, while delivering full rated output current to the external load. The slew rate is measured in terms of voltage change per unit of time. The 741 operational amplifier, for example, is rated for a slew rate of 0.5 volts per microsecond (0.5 V/μs). Slew rate is usually measured in the unity gain noninverting configuration.

Common mode rejection ratio (CMRR). A common mode voltage is one that is presented simultaneously to both inverting and noninverting inputs. In an ideal operational amplifier, the output signal resulting from the common mode voltage is zero, but in real devices it is non-zero. The common mode rejection ratio (CMRR) is the measure of the device's ability to reject common mode signals, and is expressed as the ratio of the differential gain to the common mode gain. The CMRR is usually expressed in decibels (dB), with common devices having ratings between 80 dB and 120 dB (the higher the number, the better the device).

Power supply rejection ratio (PSRR). Also called *power supply sensitivity*, the PSRR is a measure of the operational amplifier's insensitivity to changes in the power supply potentials. The PSRR is defined as the change of the input offset voltage (see below) for a one volt (1 V) change in one power supply potential (while the other is held constant). Typical values are in microvolts or millivolts per volt of power supply potential change.

Input offset voltage. The voltage required at the input to force the output voltage to zero when the input signal voltage is zero. The output voltage of an ideal operational amplifier is zero when V_{in} is zero.

Input bias current. This current is the current flowing into or out of the operational amplifier inputs. In some texts, this current is defined as the average difference between currents flowing in the inverting and noninverting inputs.

Input offset (bias) current. The difference between inverting and noninverting input bias current when the output voltage is held at zero.

Input signal voltage range. The range of permissible input voltages as measured in the common mode configuration.

Input impedance. The resistance between the inverting and noninverting inputs. This value is typically very high: 1 megohm in low-cost bipolar operational amplifiers and over 10^{12} ohms in premium BiMOS devices.

Output impedance. This parameter refers to the 'resistance looking back' into the amplifier's output terminal, and is usually modeled as a resistance between output signal source and output terminal. Typically the output impedance is less than 100 ohms.

Output short-circuit current. The current that will flow in the output terminal when the output load resistance external to the amplifier is zero ohms (i.e. a short to common).

Channel separation. This parameter is used on multiple operational amplifier integrated circuits, i.e. devices in which two or more operational amplifiers sharing the same package with common power supply terminals. The separation specification tells us something of the isolation between the op-amps inside the same package, and is measured in decibels. The 747 dual operational amplifier, for example, offers 120 dB of channel separation. From this specification we may imply that a 1 microvolt change will occur in the output voltage of one of the amplifiers when the other amplifier output voltage changes by 1 volt (the ratio 1 V/1 μV is 120 dB).

2-5.1 Minimum and maximum parameter ratings

Operational amplifiers, like all electronic components, are subject to certain maximum ratings. If these ratings are exceeded, then the user can expect either premature — often immediate — failure, or unpredictable operation. The ratings mentioned below are the most commonly used.

Maximum supply voltage. This potential is the maximum that can be applied to the operational amplifier without damaging the device. The operational amplifier uses $V+$ and $V-$ DC power supplies that are typically ±18 Vdc, although some exist with much higher maximum potentials. The maximum rating for either $V-$ or $V+$ may depend upon the value of the other (see below).

Maximum differential supply voltage. This potential is the algebraic sum of $V-$ and $V+$, namely $[(V+) - (V-)]$. It was once the case on some early devices that this rating is not the same as the summation of the maximum supply voltage ratings. For example, one 741 operational amplifier specification sheet lists $V-$ and $V+$ at 18 volts each, but the maximum differential supply voltage is only 30 volts. Thus, when both $V-$ and $V+$ are at maximum (i.e. 18 volts DC each), the actual differential supply voltage is $[(+18 \text{ V}) - (-18 \text{ V})] = 36$, which is 6 volts over the maximum rating. Therefore, when either $V-$ or $V+$ is at maximum value, the other must be proportionally lower. For example, when $V+$ is $+18$ volts, then the maximum allowable value of $V-$ is $[30 \text{ V} - 18 \text{ V}] = 12$ Vdc.

Power dissipation, P_d. This rating is the maximum power dissipation of the operational amplifier in the normal ambient temperature range ($80°C$ in commercial devices, and $125°C$ in military-grade devices). A typical rating is 500 milliwatts (0.5 watts).

Maximum power consumption. The maximum power dissipation, usually under output short circuit conditions, that the device will survive. This rating includes both internal power dissipation and device output power requirements.

Maximum input voltage. This potential is the maximum that can be applied simultaneously to both inputs. Thus, it is also the maximum common mode voltage. In most bipolar operational amplifiers the maximum input voltage is very nearly equal to the power supply voltage. There is also a maximum input voltage that can be applied to either input when the other input is grounded.

Differential input voltage. This input voltage rating is the maximum differential mode voltage that can be applied across the inverting (−IN) and noninverting (+IN) inputs.

Maximum operating temperature. The maximum temperature is the highest ambient temperature at which the device will operate according to specifications with a specified level of reliability. The usual rating for commercial devices is 70 or 80°C , while military components must operate to 125°C.

Minimum operating temperature. There is a minimum operating temperature, i.e. the lowest temperature at which the device operates within specifications. Commercial devices operate down to either 0 or −10°C , while military components operate down to −55°C.

Output short-circuit duration. This rating is the length of time the operational amplifier will safely sustain a short circuit of the output terminal. Many modern operational amplifiers are rated for indefinite output short circuit duration.

Maximum output voltage. The maximum output potential of the operational amplifier is related to the DC power supply voltages. Operational amplifiers have one or more bipolar PN junctions between the output terminal and either $V-$ or $V+$ terminals. The voltage drop across these junctions reduces the maximum achievable output voltage. For example, if there are three PN junctions between the output and power supply terminals, then the maximum output voltage is $[(V+) - (3 \times 0.7)]$, or $[(V+) - 2.1]$ volts. If the maximum $V+$ voltage permitted is 15 volts, then the maximum allowable output voltage is $[(15 \text{ V}) - (2.1 \text{ V})]$, or 12.9 volts. It is not always true, especially in older devices, that the maximum negative output voltage is equal to the maximum positive output voltage. A related rating is the maximum output voltage swing, which is the absolute value of the voltage swing from maximum negative to maximum positive.

2-6 PRACTICAL OPERATIONAL AMPLIFIERS ⎯⎯⎯⎯⎯⎯

Now that we have examined the ideal operational amplifier and some typical device specifications, let's turn our attention to practical devices. Because of its popularity and low-cost we will concentrate on the 741 device. The 741 family also includes the 747 and 1458 'dual 741' devices. Although there are many better operational amplifiers on the market, the 741 and the members of its close family are considered the industry standard generic devices.

Figure 2-6 shows the two most popular packages used for the 741. Figure 2-6A is the 8-pin miniDIP package, while Fig. 2-6B is the 8-pin

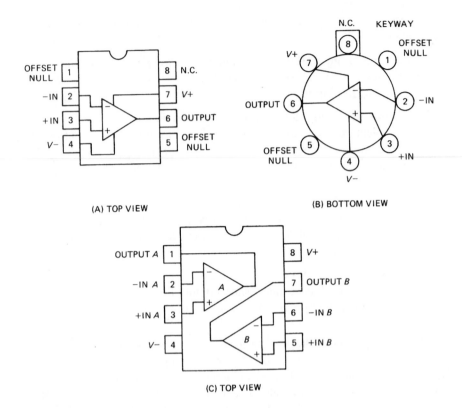

(A) TOP VIEW

(B) BOTTOM VIEW

(C) TOP VIEW

FIGURE 2-6   Typical op-amp packages: (A) 'industry-standard' 741-family devices; (B) metal can equivalent of 2-6A; (C) dual op-amp such as the LM-1458 or CA-3240.

metal can package. The 741 is also available in flat-packs and 14-pin DIP packages, although these are becoming rare. The miniDIP pinouts for a 1458 dual op-amp are shown in Fig. 2-6C. The 741 has the following pins:

−IN inverting input (Pin No. 2). The output signals produced from this input are $180°$ out of phase with the input signal applied to −IN.

+IN noninverting input (Pin No. 3). Output signals are in-phase with signals applied to the +IN input terminal.

Output (Pin No. 6). On most op-amps, the 741 included, the output is single-ended. This term means that output signals are taken between this terminal and the power supply common (see Fig. 2-7). The output of the 741 is said to be short-circuit proof because it can be shorted to common indefinitely without damage to the IC.

V+ power supply (Pin No. 7). The positive DC power supply terminal.

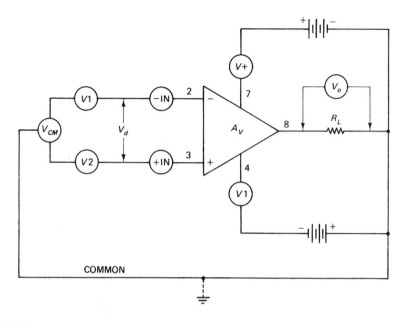

FIGURE 2-7   Equivalent circuit for signals and DC power supplies in a differential op-amp.

V− power supply (Pin No. 4). The negative DC power supply terminal.

Offset null (Pins 1 and 5). These two terminals are used to accommodate external circuitry that compensates for offset (error) voltages (see Chapter 4).

The pinout scheme shown in Fig. 2-6 is considered the *de facto* 'industry standard' for generic single operational amplifiers. Although there are numerous examples of amplifiers using different pinouts than Fig. 2-6, a very large percentage of the available devices use this scheme.

2-6.1   Standard circuit configuration

The standard circuit configuration for 741-family operational amplifiers is shown in Fig. 2-7. The pinouts are industry standard. The output signal voltage is impressed across load resistor R_L connected between the output terminal (pin no. 6) and the power supply common. Most manufacturers recommend a 2 kohm minimum value for R_L. Also, note that some operational amplifier parameters shown in Table 2-2 are based on a 10 kohm load resistance. Because it is referenced to common, the output is said to be single-ended.

The ground symbol shown in Fig. 2-7 indicates that it is optional. The point of reference for all measurements is the common connection between the two DC power supplies ($V−$ and $V+$). Whether or not this point is

TABLE 2-2 Typical op-amp parameters.

Parameter	Min.	Typical	Max.	Units
Input offset voltage V_{io}	–	1	5	mV
Input bias current	–	80	500	nA
Input resistance	0.3	2.0	–	MΩ
Input voltage range	±12	±13	–	Volts
Large signal voltage gain	50	200	–	V/mV[1]
Output voltage swing				
$R_L = 10$ K	±12	±14	–	Volts
$R_L = 2$ K	±10	±13	–	Volts
Output short circuit	–	25	–	MA
CMRR	80	90	–	dB
PSRR	77	96	–	db[2]
Risetime ($A_v = 1$)	–	0.3	–	uS
BW	–	1.16	–	MHz
Power consumption	–	50	85	mW
Power supply	–	±22	–	Vdc
Power dissipation	–	500	–	mW
Diff. input voltage	–	±30	–	Volts
Input voltage	–	±15	–	Volts

[1] $V_S = \pm15$ Vdc, $V_0 = \pm10$ V
[2] $R_L \leq 10$ KΩ

physically connected to a ground, the equipment chassis or a dedicated system ground bus is purely optional in some cases, and required in others. Whether it is required or not, however, the determination is made on the basis of circuit factors other than the basic nature of the op-amp.

The $V-$ and $V+$ DC power supplies are independent of each other. Do not make the mistake of assuming that these terminals are merely opposite ends of the same unipolar DC power supply. In fact, $V-$ is negative with respect to common, while $V+$ is positive with respect to common.

The two input signals in Fig. 2-7 are labeled $V1$ and $V2$. Signal voltage $V1$ is the single-ended potential between common and the inverting input ($-IN$), while $V2$ is the single-ended potential between common and the noninverting input ($+IN$). The 741 operational amplifier is *differential*, as indicated by the fact that both $-IN$ and $+IN$ are present. Any differential amplifier produces an output that is proportional to the difference between the two input potentials. In Fig. 2-7 the *differential input potential* V_d is the difference between $V1$ and $V2$:

$$V_d = V2 - V1 \qquad (2\text{-}5)$$

EXAMPLE 2-1

The table shown below shows six situations for input voltages to an operational amplifier. Calculate the differential input voltage, V_d, for each case.

	$V1$	$V2$	V_d
a.	+2 V	+1.5 V	?
b.	−2 V	+1.5 V	?
c.	+3 V	+5 V	?
d.	−3 V	+5 V	?
e.	+2 V	+2 V	?
f.	+1.6 V	+1.75 V	?

Solutions

(a) $V_d = V2 - V1 = (+2\ \text{V}) - (+1.5\ \text{V}) = +0.50\ \text{V}$

(b) $V_d = V2 - V1 = (-2\ \text{V}) - (+1.5\ \text{V}) = -0.50\ \text{V}$

(c) $V_d = V2 - V1 = (+3\ \text{V}) - (+5\ \text{V}) = -2\ \text{V}$

(d) $V_d = V2 - V1 = (-3\ \text{V}) - (+5\ \text{V}) = -8\ \text{V}$

(e) $V_d = V2 - V1 = (+2\ \text{V}) - (+2\ \text{V}) = 0\ \text{V}$

(f) $V_d = V2 - V1 = (+1.6\ \text{V}) - (+1.75\ \text{V}) = -0.15\ \text{V}$

Signal voltage V_{cm} in Fig. 2-7 is the *common mode* signal, that is, a potential that is common to both −IN and +IN inputs. This potential is equivalent to the situation $V1 = V2$. In an ideal operational amplifier there will be no output response at all to a common mode signal. In real devices, however, there is some small response to V_{cm}. The freedom from such responses is called the *common mode rejection ratio* (CMRR).

2-6.2   Open-loop gain (A_{vol})

The flexibility of the operational amplifier is due in large part to the extremely high open-loop DC voltage gain of the device. By definition, the open-loop voltage gain (A_{vol}) is the gain of the amplifier without feedback. If the feedback network in Fig. 2-3 had been interrupted at point X, then the gain of the circuit becomes A_{vol}. The effect of negative feedback is to reduce overall circuit gain to something less than A_{vol}.

The open-loop voltage gain of operational amplifiers is always very high. Some audio amplifier devices intended for consumer electronics applications offer A_{vol} of 20 000, while certain premium operational amplifiers offer gains to 1 000 000 and more. Depending upon the specific device surveyed, the 741 op-amp will typically exhibit A_{vol} values in the 200 000 to 300 000 range.

A consequence of such high values of A_{vol} is that very small differential input signal voltages will cause the output to saturate. On the 741 device the value of the maximum permissible output voltage, $\pm V_{sat}$, is typically about 1 volt (or a little less) below the power supply potential of the same polarity (certain BiMOS devices — such as the CA-3130 or CA-3140 — operate to within a few tenths of a volt of the supply rail). For ±15 Vdc supplies, the maximum 741 output potential is ±14 Vdc. Let's consider the maximum input potential that *will not* cause saturation of the op-amp at four popular values of power supply potential, assuming the '1 volt less' rule. The calculation is:

$$V_{in(max)} = \frac{\pm V_{sat}}{A_{vol}} \qquad (2\text{-}6)$$

$$A_{vol} = 300\,000$$

Power supply	$\pm V_{sat}$	$\pm V_{in(max)}$
$\pm$ 6 Vdc	$\pm$ 5 Vdc	17 μV
$\pm$10 Vdc	$\pm$ 9 Vdc	30 μV
$\pm$12 Vdc	$\pm$11 Vdc	37 μV
$\pm$15 Vdc	$\pm$14 Vdc	47 μV

One of the consequences of high A_{vol} is the fact that non-premium op-amps often saturate at either the $V-$ or $V+$ power supply rails when the input lines are shorted to common, and most op-amps saturate when the input lines are floating (i.e. open). This phenomena is due to tiny imbalances in the input bias conditions internal to the device, which are often said to be random in nature. Accordingly, one might expect half of a group of op-amps to saturate at $+V_{sat}$, and half to saturate at $-V_{sat}$. This situation is probably true for a very large number of devices procured from various lots and various manufacturers. However, a collection of, say, 100 devices purchased at the same time from the same manufacturer will show a marked tendency toward either $-V_{sat}$ or $+V_{sat}$, not the expected random distribution. The reason is that the input bias current imbalances tend to be design and process related, so are generally uniform from device-to-device within a given lot from the same source. For example, in a lot of 20 741 devices of the same brand tested, 19 flipped to $+V_{sat}$ and only one flipped to $-V_{sat}$ at turn-on. According to some sources we should have expected about ten to fall into each group, but that's not what usually happens in real situations.

The behavior of operational amplifiers in the open-loop configuration leads to one category of applications that takes advantage of the very high values of A_{vol}: *voltage comparators*.

2-7 VOLTAGE COMPARATORS

A voltage comparator is basically an operational amplifier that has no negative feedback network (Fig. 2-8A). The open-loop gain of the operational amplifier is very large, being on the order of 200 000 to 300 000 for many common devices. Thus, with no negative feedback the operational amplifier functions as a very high gain DC amplifier with an output that saturates at a very tiny input potential (see Example 2-1).

So what use is an amplifier that saturates with only a few microvolts of input signal voltage? Such an amplifier can be used as a voltage comparator.

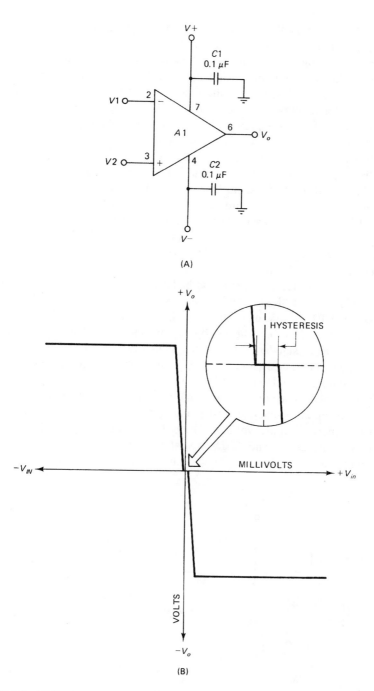

FIGURE 2-8 (A) Op-amp with DC power supply lines decoupled with bypass capacitors (C1 and C2); (B) dead-band (hysteresis) region.

The voltage comparator is used to compare two input voltages and issue an output signal that indicates their relationship ($V1 = V2$, $V1 > V2$, or $V1 < V2$). In Fig. 2-8A potential $V1$ is applied to the inverting input, and $V2$ is applied to the noninverting input. If $V1 = V2$, then $V_o = 0$. Otherwise, the output voltage obeys the relationships shown in Fig. 2-8B, which is the transfer function of the comparator. According to the normal rules for operational amplifiers, making $V1$ larger than $V2$ (see Fig. 2-8A) looks like a positive input to the inverting input, so the output potential is saturated just below $V-$. Alternatively, when $V1$ is smaller than $V2$ it looks like a negative input potential, so the output is saturated just below $V+$.

In Fig. 2-8B there is a small hysteresis band around zero where no output changes occur. This is an unfortunate defect in practical operational amplifiers. It is possible to measure the hysteresis of operational amplifiers and IC comparators in a laboratory experiment (see Section 2-10). In one such experiment involving operational amplifier and IC comparator (LM-311) devices, hysteresis bands of 1 to 25 mV were found. Not surprisingly, the low-cost 741-family of operational amplifiers had high hysteresis levels (on the order of 25 mV). The LM-311 devices had 8–10 mV hysteresis. Certain other devices had 10–20 mV of hysteresis. The overall best device in the experiment was the CA-3140, a BiMOS operational amplifier. The CA-3140 device uses the industry standard 741 pinouts, which are shown in Fig. 2-8A.

The LM-311 device (Fig. 2-9A) is a low-cost voltage comparator in IC form. Although based internally on op-amp circuitry, this device is specifically designed as a voltage comparator. It has a ground terminal (pin 1), contrary to op-amp practice, and requires an output pull-up resistor (R) to a positive power supply voltage. The output terminal can drive loads such as relay coils, lamps and LEDs operated at potentials up to 40–50 volts (depending upon category of device) and current loads of 50 milliamperes.

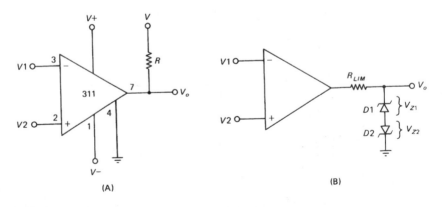

FIGURE 2-9 (A) Normal connections for the LM-311 comparator IC; (B) Output sharpening using zener diodes.

If the LM-311 is operated to be compatible with TTL digital logic, then the pull-up resistor is terminated in a +5 VDC potential, and usually has a value of 1.5 to 3.3 kohms.

A means for limiting the output level, improving the sharpness of the transfer function corners (see Fig. 2-8B), and improving speed by reducing latch-up problems, is shown in Fig. 2-9B. In this circuit a pair of back-to-back zener diodes are connected across the output line. When the output voltage is HIGH it is limited to $V_{Z1} + 0.7$ volts; and when LOW it is limited to $V_{Z2} + 0.7$ volts. These potentials represent the reverse bias zener voltages of $D1$ and $D2$, plus the normal forward bias voltage drop of the other diode (which is forward biased).

2-7.1 High drive capacity comparators

Figure 2-10 shows a means for increasing the drive capacity of the comparator. In this circuit a bipolar transistor (2N3704, 2N2222, etc.) is used to control a larger load than the device could normally handle (such as the relay coil shown here). The output voltage (V_o) of the comparator is used to set up the DC bias for the NPN transistor. When the comparator output is HIGH, then the transistor is biased hard-on and the load is grounded through the transistor's collector–emitter path. Alternatively, when the comparator output is LOW, then the transistor is reverse biased and the load remains ungrounded.

The diode across the relay coil is essential for any inductive load. When the magnetic field surrounding an inductor (such as a relay coil) collapses the counterelectromotive force (CEMF) generates a high-voltage spike that is capable of damaging components or interrupting circuit operation (especially digital circuits). The diode is normally reverse biased, but for the CEMF

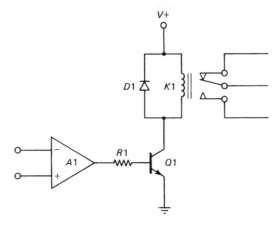

FIGURE 2-10 Using an external transistor to increase load capacity, as in the relay coil driver shown here.

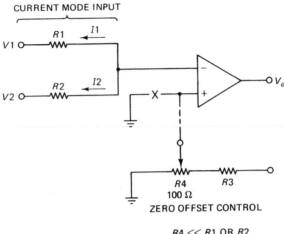

FIGURE 2-11   Current mode comparator operation.

spike it is forward biased. The diode therefore clamps the high voltage spike to about 0.7 volts.

2-7.2   Current mode comparator

Figure 2-11 shows two comparator circuit techniques applied to the same circuit. One technique is a *zero offset control* used to reduce the effects of the hysteresis band, while the other technique is the *current mode* configuration. The offset control (R4) biases one input to a small but non-zero level so that it is ready to trip when the other input is also non-zero. In this particular case the noninverting input is grounded (V2 = 0), but could as easily be connected to a non-zero voltage.

Current mode operation is usually faster and less prone to latch-up than voltage mode operation. For this reason, current mode comparators are sometimes used in high speed analog to digital converters (A/D). Assume that the noninverting input is grounded. In this case, the output potential V_o will reflect the relationship of the two currents. If $I1 = I2$, then $V_o = 0$. This circuit is, to the outside observer, a voltage comparator in that $I1 = V1/R1$ and $I2 = V2/R2$. Of course, the circuit is also useful for accommodating current output devices such as the LM-334 temperature monitor IC as well as voltages as shown.

2-7.3   Zero-crossing detectors

Figure 2-12A shows a *zero-crossing detector circuit*. In this case a comparator is connected with its noninverting input grounded. When V_{in} is non-zero, then the output will also be non-zero. But when the input voltage crosses zero, the

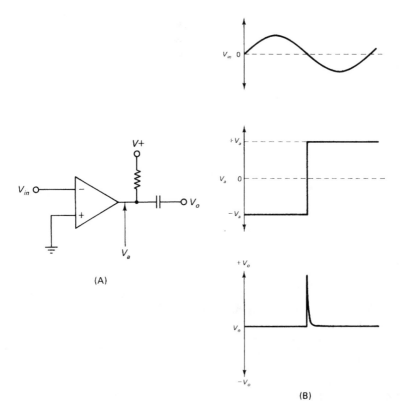

FIGURE 2-12 (A) Zero-crossing detector circuit; (B) waveforms (top is input, middle is output of op-amp, and bottom is output from capacitor).

output briefly goes to zero, producing the differentiated output pulse shown. These relationships are shown in Fig. 2-12B.

2-7.4 Window comparators

A *window comparator* is shown in Fig. 2-13. This circuit consists of two voltage comparators connected such that one or other input is activated when the input voltage (V_{in}) exceeds either positive or negative limits. The limits are set by setting $V1$ or $V2$ reference voltages. A possible application for this circuit is alarm systems (for example over- and under-temperature alarms), and other applications where a range of permissible values exists between two forbidden regions.

2-7.5 Pre-biased comparator (voltage level detector)

Figure 2-14A shows a method for biasing either comparator input to a specific reference voltage. This circuit is called a *voltage level detector*. Although in

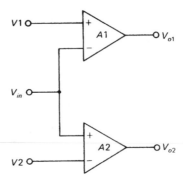

FIGURE 2-13   Basic window comparator circuit.

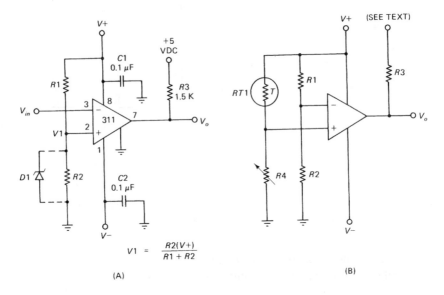

$$V1 = \frac{R2(V+)}{R1 + R2}$$

(A)

(B)

FIGURE 2-14   (A) LM-311 comparator with (+) input biased with zener diode; (B) temperature alarm circuit based on comparator.

this case the noninverting input is biased and the inverting input is active, the roles can just as easily be reversed. Two methods of biasing are used: resistor voltage divider and zener diode. If $R2$ is replaced with a zener diode, then the reference potential is the zener potential. In that case, $R1$ is the normal current limiting resistor needed to protect the zener from self-destruction. In the case where a resistor voltage divider is used, the bias voltage $V1$ is set by the voltage divider equation:

$$V1 = \frac{(V+)R2}{R1 + R2} \qquad (2\text{-}7)$$

For example, suppose $R1 = R2 = 10$ kohms, and $V+ = 12$ VDC:

$$V1 = \frac{(V+)R2}{R1 + R2}$$

$$V1 = \frac{(+12 \text{ VDC})(10 \text{ kohms})}{(10 \text{ kohms} + 10 \text{ kohms})}$$

$$V1 = \frac{120 \text{ volts}}{20} = 6 \text{ volts}$$

2-7.6 Temperature alarm

Figure 2-14B shows an over-temperature circuit based on Fig. 2-14A. In this circuit the inverting input is biased by voltage divider $R1/R2$, while the noninverting input is set by another voltage divider, $R4/RT1$. Resistance $RT1$ is a *thermistor*, which has a resistance proportional (or inversely proportional in some types) to the temperature. Potentiometer $R4$ is used to set the 'trip-point' temperature. The values of the resistors depends upon the set trip-point desired and the resistance of the thermistor over the range of temperatures being monitored.

2-7.7 Pulse width controller

Pulse width modulation is used in many communications systems, motor and load controllers, switching DC power supplies and other applications. A pulse width modulator will vary the width of an output pulse proportionally to an applied input voltage. We can use a voltage comparator to form a basic pulse width modulator, as shown in Fig. 2-15A. In this circuit a triangle waveform is applied to the inverting input of the comparator, while the DC reference potential (V_{ref}) output of a potentiometer is applied to the noninverting input.

Figures 2-15B and 2-15C show the relationship between the output terminal and the two input signals, V_a and V_{ref}. Examine Fig. 2-15B. Note that V_{ref} is a positive DC potential. As long as the applied triangle waveform is less than $+V_{ref}$, the output of the comparator, V_o, is HIGH. But at time $T1$ the two voltages become equal, so the output flips as the triangle waveform increases in value above $+V_{ref}$. During the period $T2$–$T1$ signal V_a is greater than $+V_{ref}$, so V_o is LOW. At time $T2$, however, the situation again reverts back to the situation in which V_a is less than $+V_{ref}$ so V_o drops LOW again.

Now consider a slightly different situation in Fig. 2-15C. In this case the value of V_{ref} is readjusted to a negative value, $-V_{ref}$. As in the previous case, the output V_o is LOW during the period $T2 - T1$, but note that this segment is much longer than it was in Fig. 2-15B. This difference is caused by the relationship between V_{ref} and V_a under the two different situations.

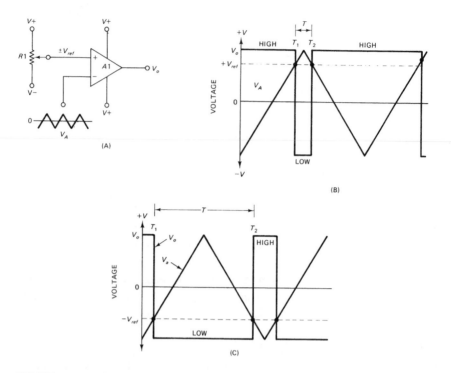

FIGURE 2-15 (A) Pulse width controller; (B) waveform with high $+V_{ref}$ setting of $R1$; (C) with high $-V_{ref}$ setting of $R1$.

2-7.8 Analog-to-digital converter

An *analog-to-digital converter* (A/D or ADC) is a circuit that converts an analog voltage into a binary number that is proportional to that voltage. For example, in an eight-bit unipolar ADC we might find that binary number 00000000 represents zero-volts, while 11111111 represents the full-scale potential (e.g. 2.56 volts, 5 volts, 10 volts). A digital-to-analog converter (DAC) is exactly the opposite: it converts a binary number into an analog voltage. In both cases, the minimum step of output change is called the 'one least significant bit' (1-LSB) value, i.e. the amount of change caused by a change in the least significant bit of the binary number (e.g. xxxxxxx0 to xxxxxxx1, or vice versa). Figure 2-16 shows one type of ADC circuit that can be implemented with a voltage comparator and a DAC circuit. This type of ADC circuit is known both as the *binary ramp* ADC and the *servo ADC*.

The comparator noninverting input in the circuit of Fig. 2-16A is connected to the unknown voltage, V_x, while the inverting input is connected to the output of the DAC, V_a. The output state of the comparator indicates the relationship of V_a and V_x, that is: $V_a < V_x$ or $V_a = V_x$. These relationships are graphed in Fig. 2-16B. The DAC output voltage is a stepped potential that is

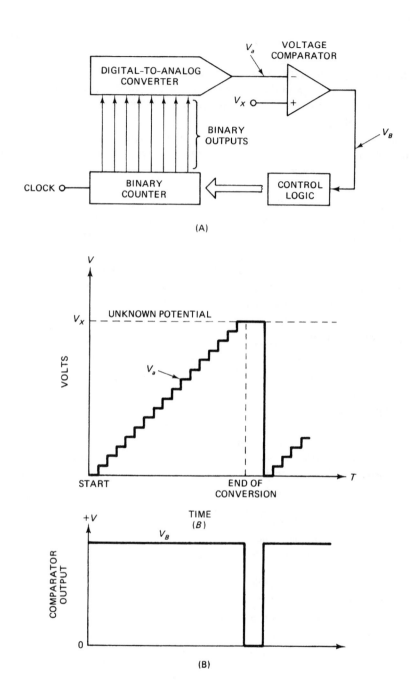

FIGURE 2-16 (A) Analog-to-digital converter circuit; (B) timing waveforms.

raised by the 1-LSB amount every time the binary counter is clocked. During this time the comparator output (V_b) is HIGH. At the point where the DAC output voltage equals (or exceeds) the unknown input voltage, the comparator output switches to the LOW condition, indicating the end of conversion. The binary number appearing on the output of the binary counter represents the unknown voltage.

A variant on this type of circuit is often used in microprocessor-based instrumentation. The binary counter is replaced with an eight-bit output port on the computer, while the comparator output is read by one bit of an input port. Software replaces the control logic. The program will output a binary number starting with 00000000, test the comparator output for HIGH or LOW, and then branch according to the results of the test. For example, if the output is HIGH, it will increment the binary number and output this new number to the DAC. This process continues in an iterative fashion until the comparator output drops LOW indicating the condition $V_a = V_x$. At this point the computer program branches to store the binary number in a memory location, or use it in a calculation.

2-8 SUMMARY

1. Operational amplifiers were initially designed to perform mathematical operations in analog computers. Because of this fact op-amps use bipolar power supplies to accommodate both polarity outputs, and need both inverting and noninverting inputs. Most op-amps are differential amplifiers.

2. There are a number of categories of operational amplifiers: general purpose, voltage comparators, low input current, low noise, low power, low drift, wide bandwidth, single DC supply, high voltage, multiple devices, limited/special purpose, and instrumentation amplifiers.

3. The ideal operational amplifiers has seven properties:
   (a) infinite open-loop voltage gain;
   (b) infinite input impedance;
   (c) zero output impedance;
   (d) zero noise contribution;
   (e) zero DC output offset;
   (f) infinite bandwidth; and
   (g) both differential inputs stick together.

4. There are two configurations for the operational amplifier: inverting follower and noninverting follower.

5. The differential input voltage is the difference between single-ended voltages applied to the $-$IN ($V1$) and $+$IN ($V2$) inputs: $V_d = (V2 - V1)$.

6. A common mode voltage (V_{cm}) is one that applies equally to both $-$IN and $+$IN inputs. It is equivalent to the situation where $V1 = V2$. The common mode rejection ratio (CMRR) is a measure of the ability of the differential amplifier to reject common mode voltages.

7. Open-loop voltage gains range from 20 000 to more than 1 000 000, with 200 000 to 300 000 being common in 741 devices.

8. A voltage comparator is an operational amplifier with no feedback. It compares the two input voltages, and issues an output signal that indicates whether $V1 = V2$, $V1 > V2$ or $V1 < V2$.

2-9 RECAPITULATION

Now return to the objectives and Pre-quiz questions at the beginning of the chapter and see how well you can answer them. If you cannot answer certain questions, place a check mark to each and review the appropriate parts of the text. Next, try to answer the questions and work the problems below, using the same procedure.

2-10 LABORATORY EXERCISES

1. Connect the circuit of Fig. 2-E1, using a 741 operational amplifier. The output indicator is a digital voltmeter. (a) ground both $-$IN and $+$IN inputs (pins 2 and 3). Note what happens to the output voltage when $V-$ and $V+$ are applied simultaneously; (b) repeat the experience using a large variety of 741 devices from different manufacturers; (c) repeat the experiment with $-$IN and $+$IN floating (note: break the circuit at points X).

2. Connect the circuit of Fig. 2-E2, initially with the inverting input connected to the wiper terminal of potentiometer $R1$. Connect the noninverting input to a positive voltage between $+1$ Vdc and $+10$ Vdc (the $+5$ Vdc shown in Fig. 2-E2 was used because it is commonly available on laboratory breadboard power supplies): (a) starting with point B at zero volts, slowly increase the potential by adjusting $R1$ until a transition of output potential at point C occurs. Measure and record

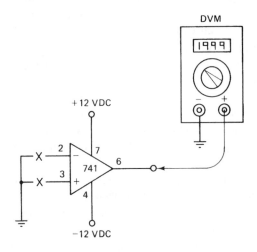

FIGURE 2-E1

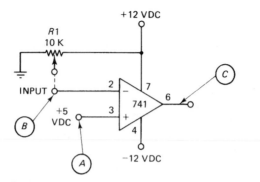

FIGURE 2-E2

voltages present at A and B; (b) repeat the experiment starting with the voltage at point A at +12 Vdc; (c) disconnect the inverting input from R1, and reconnect to the output of a function generator (monitor point C with an oscilloscope and note the relationship between input and output signals.

2-11 QUESTIONS AND PROBLEMS

1. List several areas where operational amplifiers are typically found.

2. List the three major sections of the internal circuitry of the operational amplifier.

3. A DC differential amplifier has two inputs: _____ , which produces out of phase output signals, and _____ which produces in-phase output signals.

4. List five different categories of operational amplifiers.

5. List the seven properties of the ideal operational amplifier.

6. The open-loop gain of an operational amplifier is 300 000. Calculate the closed-loop gain if the feedback factor (B) is 0.01.

7. A potential of +5 Vdc is applied to the noninverting input of an operational amplifier. A voltmeter connected between the inverting input and common will read: _____ Vdc.

8. The _____ _____ rejection ratio relates output change to a change in either $V-$ or $V+$, while the other is held constant.

9. A dual operational amplifier has a channel separation of 100 dB. Calculate the output voltage change at amplifier B if the output of amplifier A changes 2.5 volts.

10. The 'industry standard' operational amplifier pinouts are the same pinouts as found on the _____ operational amplifier.

11. There is no _____ terminal on the operational amplifier. Output signal is taken between the output terminal and the power supply _____ line.

12. Calculate the differential input voltage (V_d) if an operational amplifier has 1.83 Vdc applied to the $-$IN input, and 4.5 Vdc applied to the $+$IN input.

13. Calculate the differential input voltage (V_d) if an operational amplifier has 1.05 Vdc applied to the $-$IN input, and -2.5 Vdc applied to the $+$IN input.

14. Calculate the input voltage that will cause saturation at the positive output limit if $+V_{sat} = 13.4$ Vdc, and $A_{vol} = 250\,000$.

15. Calculate the input voltage that will cause saturation at the negative output limit if $-V_{sat} = -14$ Vdc, and $A_{vol} = 300\,000$.

16. A constant current source used in a transistor DC differential amplifier provides 12 mA of current. If $I1 = 4.2$ mA, then $I2 = $ _____ mA.

17. A Darlington input amplifier is characteristic of a _____ operational amplifier.

18. _____ _____ operational amplifiers are internally compensated to minimize thermal variations of the output voltage.

19. A DC voltmeter connected to the noninverting input of an operational amplifier measures $+4.5$ Vdc. What is the DC voltage present at the inverting input if $V_{in} = 0$ volts?

20. An operational amplifier has an open-loop voltage gain of $500\,000$. Calculate the closed-loop voltage gain of a simple resistor voltage divider consisting of a 120 kohm resistor from the output of the inverting input, and a 27 kohm resistor from the inverting input to ground.

21. Calculate the noise power dissipated in a 2.2 megohm resistor that is operated at a temperature of $45°C$. Assume a bandwidth of 100 kHz.

22. Calculate the noise voltage across the resistor in question 21 above.

23. Calculate the noise voltage present across a 1.5 megohm resistor operated at room temperature ($25°C$) when a 100 pF capacitor is shunted across the resistor.

24. Define *virtual ground* in your own words.

25. Define *slew rate* and state the units normally used to measure slew rate.

26. Define *common mode rejection ratio* (CMRR).

27. Define *power supply rejection ratio* (PSRR).

28. Draw the circuit diagram for an operational amplifier operated as a voltage comparator.

29. Draw the circuit diagram for a current-mode comparator based on an operational amplifier.

30. Draw the circuit of a voltage comparator used as a voltage-crossing detector.

31. Draw the circuit diagram for an operational amplifier window detector.

32. Draw the circuit for a pre-biased comparator used as a voltage level detector.

CHAPTER 3

Inverting and noninverting operational amplifier configurations

OBJECTIVES

1. Learn to recognize the three principal operational amplifier configurations.
2. Be able to set the gain of either inverting or noninverting amplifiers, and use them in practical applications.
3. Learn the advantages and disadvantages of the principal op-amp circuit configurations.
4. Be able to describe the response of linear amplifiers to common DC, AC and composite AC/DC signals.

3-1 PRE-QUIZ

These questions test your prior knowledge of the material in this chapter. Try answering them before you read the chapter. Look for the answers (especially those you answered incorrectly) as you read the text. After you have finished studying the chapter try answering these questions again, and those at the end of the chapter (see Section 3-8).

1. An inverting follower must have a voltage gain of -10, and exhibit an input impedance of not less than 22 kohms. Find the value of feedback resistor R_f that will allow the required gain with the specified minimum value of input impedance.

2. An inverting amplifier has an input resistor of 10 kohms, and a feedback resistor of 1 megohm. Find the output voltage if an input signal of $+100$ mVdc is applied to the input.

3. Calculate the value of capacitor needed in the amplifier described in the previous example to shunt R_f that would limit the upper AC frequency response to 100 Hz.

4. A noninverting follower uses a 1 kohm 'input' resistor and a 100 kohm feedback resistor. What is the voltage gain of this circuit?

3-2 INVERTING FOLLOWER CIRCUITS

The *inverting follower* is an operational amplifier circuit configuration in which the output signal is 180° out of phase with the input signal. Figure 3-1A is a cathode ray oscilloscope (CRO) presentation that shows the relationship

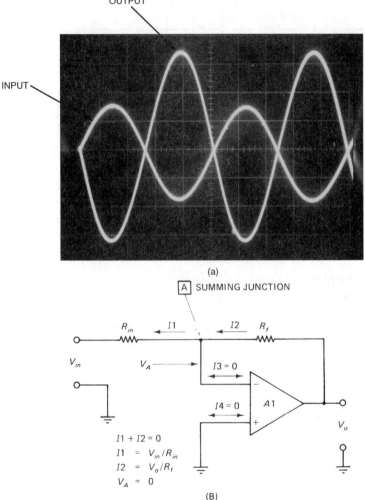

FIGURE 3-1   (A) Input and output waveforms for inverting follower; (B) inverting follower circuit; (C) transfer function.

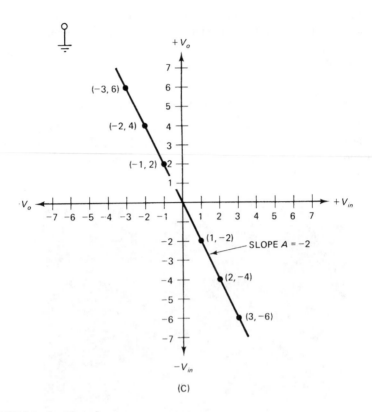

FIGURE 3-1 (continued)

between input and output signals for an inverting follower with a gain of −2. Note the phase reversal present in the output signal with respect to the input signal. In order to achieve this inversion, the inverting input (−IN) of the operational amplifier is active, and the noninverting input (+IN) is grounded.

Figure 3-1B shows the basic configuration for inverting follower (also called inverting amplifier) circuits. The noninverting input is not used, so is set to ground potential. There are two resistors in this circuit: resistor R_f is the negative feedback path from the output to the inverting input, while R_{in} is the input resistor. We will examine the R_f/R_{in} relationship in order to determine how gain is fixed in this type of circuit. But first, let's take a look at the implications of grounding the noninverting input in this type of circuit.

3-2.1 What is a `virtual ground´?

A *virtual ground* is a ground that only acts like a ground, but might not have a physical connection to ground. While this definition sounds strange at first, it's not an unreasonable description of a virtual ground. Unfortunately, that terminology is confusing and therefore leads to an erroneous implication that the virtual ground somehow doesn't really function as a ground. Let's examine the concept of a virtual ground.

In Chapter 2 you learned the properties of the 'Ideal Operational Amplifier'. One of those properties tells us that differential inputs 'stick together'. Put another way, this property means that a voltage applied to one input appears on the other input also. In the arithmetic of op-amps, therefore, we must *treat both inputs as if they are at the same potential*. This fact is not merely a theoretical device, either, for if you actually apply a potential, say 1 Vdc, to the noninverting input the same 1 Vdc potential can also be measured at the inverting input.

In Fig. 3-1A, the noninverting input is grounded, so it is at 0 Vdc potential. This fact, by the properties of the ideal op-amp, means that the inverting input of the op-amp is also at the same 0 Vdc ground potential. Since the inverting input is at ground potential, but has no physical ground connection, it is said to be at 'virtual' (as opposed to 'physical') ground. A virtual ground is, therefore, *a point that is fixed at ground potential (0 Vdc), even though it is not physically connected to the actual ground or common of the circuit.* The choice of the term virtual ground was unfortunate, for the concept is actually quite simple even though the terminology makes it sound a lot more abstract than it really is.

3-2.2 Developing the transfer equation for the inverting amplifier

The transfer equation of any circuit is the output function divided by the input function. For an operational amplifier used as a voltage amplifier, therefore, the transfer function describes the voltage gain:

$$A_v = \frac{V_o}{V_{in}} \tag{3-1}$$

where:
  A_v is the voltage gain (dimensionless)
  V_o is the output signal potential
  V_{in} is the input signal potential

(Note: V_o and V_{in} are in the same units.)

EXAMPLE 3-1

What is the voltage gain of an amplifier if +100 mVdc (0.100 Vdc) at the input produces a −10 Vdc output potential?

Solution

$$A_v = \frac{V_o}{V_{in}}$$

$$A_v = \frac{-10 \text{ V}}{0.1 \text{ V}} = -100 \qquad \blacksquare$$

In the inverting follower circuit (Fig. 3-1A) the gain is set by the ratio of two resistors, R_f and R_{in}. Let's make a step-by-step analysis to see if we can develop this relationship. Consider the currents flowing in Fig. 3-1B. The input bias currents, I_3 and I_4, are assumed to be zero for purposes of this analysis. This is a reasonable assumption because our model is an *ideal* operational amplifier. In a real op-amp these currents are non-zero and have to be accounted for, but here we use the ideal model. Thus, in the analysis below we can ignore bias currents (assume that $I_3 = I_4 = 0$).

Remember that the *summing junction* (point A) is a virtual ground and is at ground potential because the noninverting input is grounded. Current I_1 is a function of the applied input voltage, V_{in}, and the input resistance R_{in}. By Ohm's law, then, the value of I_1 is:

$$I_1 = \frac{V_{in}}{R_{in}} \qquad (3\text{-}2)$$

Further, we know that current I_2 is also related by Ohm's law to the output voltage, V_o, and the feedback resistor R_f (again, because the summing junction is at 0 Vdc):

$$I2 = \frac{V_{in}}{R_f} \qquad (3\text{-}3)$$

How are I_1 and I_2 related? These two currents are the only currents entering or leaving the summing junction (recall that $I_3 = 0$), so by *Kirchhoff's current law* (KCL) we know that:

$$I_1 + I_2 = 0 \qquad (3\text{-}4\text{A})$$

so, therefore

$$I_2 = -I_1 \qquad (3\text{-}4\text{B})$$

We can arrive at the transfer function by substituting Eqs (3-2) and (3-3) into Eq. (3-4B):

$$I_2 = -I_1 \qquad (3\text{-}4\text{B restated})$$

$$\frac{V_o}{R_f} = -\frac{V_{in}}{R_{in}} \qquad (3\text{-}5)$$

Algebraically rearranging Eq. (3-5) yields the transfer equation in standard format:

$$\frac{V_o}{V_{in}} = \frac{-R_f}{R_{in}} \qquad (3\text{-}6)$$

According to Eq. (3-1), the gain (A_v) of the circuit is V_o/V_{in}, so we may also write Eq. (3-6) in the form:

$$A_v = \frac{-R_f}{R_{in}} \quad (3\text{-}7)$$

EXAMPLE 3-2

What is the gain of an op-amp inverting follower if the input resistor (R_{in}) is 10 kohms, and the feedback (R_f) resistor is 1 megohm?

Solution

$$A_v = \frac{-R_f}{R_{in}}$$

$$A_v = \frac{-1\,000\,000 \ \Omega}{10\,000 \ \Omega} = -100 \qquad \blacksquare$$

We have shown above that the voltage gain of an op-amp inverting follower is merely the ratio of the feedback resistance to the input resistance ($-R_f/R_{in}$). The minus sign indicates that a 180° phase reversal takes place. Thus, a negative input voltage produces a positive output voltage, and vice versa.

We often see the transfer equation (Eq. (3-6)) written to express output voltage in terms of gain and input signal voltage. The two expressions are:

$$V_o = A_v V_{in} \quad (3\text{-}8)$$

and

$$V_o = -V_{in} \left(\frac{R_f}{R_{in}} \right) \quad (3\text{-}9)$$

EXAMPLE 3-3

What is the output voltage from an inverting follower in which $V_{in} = 100$ mV (0.100 V), $R_f = 100$ kohms and $R_{in} = 2$ kohms?

Solution

$$V_o = -V_{in} \left(\frac{R_f}{R_{in}} \right)$$

$$V_o = -(0.100) \left(\frac{100 \ k\Omega}{2 \ k\Omega} \right) = -5 \text{ volts} \qquad \blacksquare$$

In the example above the voltage gain (A_v) is $-R_f/R_{in} = -100$ kohms/2 kohms $= -50$.

The transfer function ($A_v = V_o/V_{in}$) can be plotted on graph paper in terms of input and output voltage. Figure 3-1C shows the plot V_o versus V_{in} for an inverting amplifier with a gain of -2. Straight lines plotted as in Fig. 3-1C have the mathematical form $Y = aX + b$, where a is the slope of the line, and b is the Y-intercept. In the case of a perfect amplifier, the Y-intercept is 0 volts. Given the nature of Fig. 3-1C the basic form becomes for our purposes $V_o = A_v V_{in} \pm V_{offset}$.

3-2.3 Inverting amplifier transfer equation by feedback analysis

In Section 3-2.2 we developed the inverting amplifier transfer equation from the ideal model of the operational amplifier. Now let's consider the same matter from the point of view of the generic feedback amplifier to see if Eq. (3-7) is valid. When used in a closed-loop circuit the operational amplifier is a feedback amplifier, so feedback analysis will result in the same transfer equation as the 'ideal model' analysis.

Figure 3-2 shows an operational amplifier with its feedback network. The overall gain of this type of amplifier is defined by the following expression:

$$A_v = \frac{-A_{vol}C}{1 + A_{vol}\beta} \tag{3-10}$$

where:

A_v is the closed-loop voltage gain
A_{vol} is the open-loop voltage gain
C is the transfer equation of the input network
β is the transfer equation of the feedback network

There are two networks that must be considered in this analysis: the input network (C) and the feedback network (β); both networks are resistor voltage

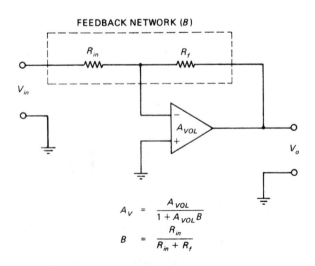

FIGURE 3-2   Model of inverting follower showing feedback network and forward again A_{vol}.

divider attenuators so we can expect β and C to be fractions. The expression for the input network C in Fig. 3-2 is:

$$C = \frac{R_f}{R_f + R_{in}} \qquad (3\text{-}11)$$

The C term is needed because the input signal is attenuated by the R_{in}/R_f voltage divider network. If the signal is applied directly to the inverting input, as it might be in certain other feedback amplifiers, then this input attenuation term is unity, so it disappears from Eq. (3-10).

The feedback transfer equation is defined by the feedback voltage divider, R_f/R_{in}:

$$\beta = \frac{R_{in}}{R_f + R_{in}} \qquad (3\text{-}12)$$

We can now substitute the expressions for β (Eq. (3-12)) and C (Eq. (3-11)) into the equation for the standard feedback amplifier (Eq. (3-10)):

$$A_v = \frac{-A_{vol}C}{1 + A_{vol}\beta}$$

therefore,

$$A_v = \frac{-A_{vol}\left(\dfrac{R_f}{R_f + R_{in}}\right)}{1 + A_{vol}\left(\dfrac{R_{in}}{R_f + R_{in}}\right)} \qquad (3\text{-}13)$$

We may legally divide both numerator and denominator by the open-loop gain, which yields:

$$A_v = \frac{-\left(\dfrac{A_{vol}}{A_{vol}}\right)\left(\dfrac{R_f}{R_f + R_{in}}\right)}{\left(\dfrac{1}{A_{vol}}\right) + \left(\dfrac{R_{in}}{R_f + R_{in}}\right)} \qquad (3\text{-}14)$$

Because A_{vol} is infinite in ideal devices (and very, very high in practical devices), the term $1/A_{vol} \rightarrow 0$, so we may write Eq. (3-14) in the form:

$$A_v = \frac{-\left(\dfrac{R_f}{R_f + R_{in}}\right)}{\left(\dfrac{R_{in}}{R_f + R_{in}}\right)} \qquad (3\text{-}15)$$

Earlier we discovered that $A_v = -R_f/R_{in}$. If the feedback analysis is correct, then Eq. (3-15) will be equal to $-R_f/R_{in}$. Solving this relationship we invert and multiply:

$$\left(\frac{R_f + R_{in}}{R_{in}}\right) \times \left(\frac{-R_f}{R_f + R_{in}}\right) = \frac{-R_f}{R_{in}} \tag{3-16}$$

$$\frac{R_f}{R_{in}} = \frac{R_f}{R_{in}} \tag{3-17}$$

Equation (3-17) demonstrates the equality of the two methods, proving that the transfer equation (Eq. (3-7)) derived earlier is valid.

The following equations apply to inverting followers:

$$A_o = \frac{-R_f}{R_{in}} \tag{3-18}$$

$$V_o = -A_v V_{in} \tag{3-19}$$

$$V_o = -V_{in}\left(\frac{R_f}{R_{in}}\right) \tag{3-20}$$

3-2.4 Multiple input inverting followers

We can accommodate multiple signal inputs on an inverting follower by using a circuit such as Fig. 3-3. There are a number of applications of such

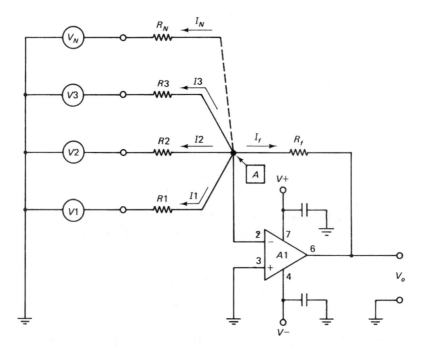

FIGURE 3-3   Multiple input inverting follower.

circuits: summers, audio mixers, instrumentation and so forth. The multiple input inverter of Fig. 3-3 can be evaluated exactly like Fig. 3-1B, except that we have to account for more than one input. Again appealing to KCL, we know that:

$$I_1 + I_2 + I_3 + \cdots + I_n = -I_f \tag{3-21}$$

Also by Ohm's law, considering that summing junction A is virtually grounded, we know that:

$$I_1 = \frac{V_1}{R_1} \tag{3-22}$$

$$I_2 = \frac{V_2}{R_2} \tag{3-23}$$

$$I_3 = \frac{V_3}{R_3} \tag{3-24}$$

$$I_n = \frac{V_n}{R_n} \tag{3-25}$$

$$I_f = \frac{V_o}{R_f} \tag{3-26}$$

Substituting Eqs (3-22) through (3-26) into Eq. (3-21):

$$\frac{V_1}{R_1} + \frac{V_2}{R_2} + \frac{V_3}{R_3} + \cdots + \frac{V_n}{R_n} = \frac{-V_o}{R_f} \tag{3-27}$$

or, algebraically re-arranging Eq. (3-27) to solve for V_o:

$$V_o = -R_f \left[\frac{V_1}{R_1} + \frac{V_2}{R_2} + \frac{V_3}{R_3} + \cdots + \frac{V_n}{R_n} \right] \tag{3-28}$$

Equation (3-28) is the transfer equation for the multiple input inverting follower. The terms V_n and R_n refer to the nth voltage and nth resistance, respectively. Let's consider a three input example ($n = 3$).

EXAMPLE 3-4

A circuit such as Fig. 3-3 has a 100 kohm feedback resistor (R_f), and the following input resistors: $R_1 = 10$ kohms, $R_2 = 50$ kohms, and $R_3 = 100$ kohms. Find the output voltage when the following input voltages are present: $V_1 = 0.100$ volts, $V_2 = 0.200$ volts and $V_3 = 1$ volt.

Solution

$$V_o = -R_f \left[\frac{V_1}{R_1} + \frac{V_2}{R_2} + \frac{V_3}{R_3} \right]$$

$$V_o = (-100 \text{ k}\Omega) \left[\frac{0.100 \text{ V}}{10 \text{ k}\Omega} + \frac{0.200 \text{ V}}{50 \text{ k}\Omega} + \frac{1.00 \text{ V}}{100 \text{ k}\Omega} \right]$$

$$V_o = -([(10)(0.100)] + [(2)(0.200)] + [(1)(1.00)]) \text{ volts}$$

$$V_o = -[(1) + (0.4) + (1)] \text{ volts} = 2.4 \text{ volts} \qquad \blacksquare$$

3-2.5 Output current

The output current I_o must be supplied by the output terminal of the op-amp. Typically small-signal op-amps supply 5 to 25 mA of current depending on the device, while power op-amps such as the Burr-Brown OPA-511 supplies up to 5 A at potentials of ±30 volts. The output current (I_o) splits into two paths (Fig. 3-4); a portion of the output current flows into the feedback path (I_f) and a portion flows into the load (I_L). The total current is:

$$I_o = I_f + I_L \qquad (3\text{-}29)$$

where:

I_o is the output current
I_f is the feedback current (V_o/R_f)
I_L is the load current (V_o/R_L)

In normal voltage amplifier service both I_f and I_L tend to be very small compared with the available output current. But in applications where load and feedback resistances are low, then the output currents may approach the maximum specified value. To determine whether this limit is exceeded divide output potential V_o by the parallel combination of R_f and R_L:

$$I_{MAX} = \frac{V_{o(max)}(R_f + R_L)}{R_f R_L} \qquad (3\text{-}30)$$

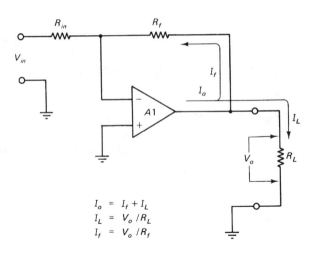

FIGURE 3-4 Output current relations in inverting follower.

where:

$V_{o(max)}$ is the maximum expected output potential in volts (V)
I_{max} is the maximum allowable output current in amperes (A)
R_f is the feedback resistance in ohms (Ω)
R_L is the load resistance in ohms (Ω)

In general, the output current limit is not approached on ordinary devices unless load or feedback resistances are less than 1000 ohms. Power devices, of course, can drive a lower load or feedback resistance combination.

3-2.6 Response to AC signals

Thus far our discussion of inverting amplifiers has assumed a DC input signal voltage. The behavior of the circuit in response to AC signals (e.g. sinewaves, squarewaves and triangle waves) is similar. Recall the rules for the inverter: positive input signals produce negative output signals, and negative input signals produce positive output signals. These relationships mean that a 180° phase shift occurs between input and output. The relationship is shown in Fig. 3-5A.

Although the DC-coupled op-amp will respond to AC signals, there is a limit that must be recognized. If the peak value of the input signal becomes too great, then output clipping (Fig. 3-5B) will occur. The peak output voltage will be:

$$V_{o(peak)} = A_v V_{in(peak)} \qquad (3\text{-}31)$$

where:

$V_{o(peak)}$ is the peak output voltage
$V_{in(peak)}$ is the peak input voltage
A_v is the voltage gain

For every value of the $V-$ and $V+$ power supply potentials there is a maximum attainable output voltage, $V_{o(max)}$. As long as the peak output voltage is less than this maximum allowable output potential, then the input waveform will be faithfully reproduced in the output (except amplified and inverted). But if the value of $V_{o(peak)}$ determined by Eq. (3-31) is greater than $V_{o(max)}$, then clipping will occur.

In a linear voltage amplifier clipping is undesirable. The maximum output voltage can be used to calculate the maximum input signal voltage:

$$V_{in(max)} = \frac{V_{o(max)}}{A_v} \qquad (3\text{-}32)$$

There are occasions when clipping is desired. For example, in radio transmitter circuits called modulation limiters are often simple clippers followed by an audio low-pass filter that removes the harmonic distortion created by clipping. Another case where clipping is desired is in generating squarewaves from sinewaves. The goal in that case is to drive the input so hard

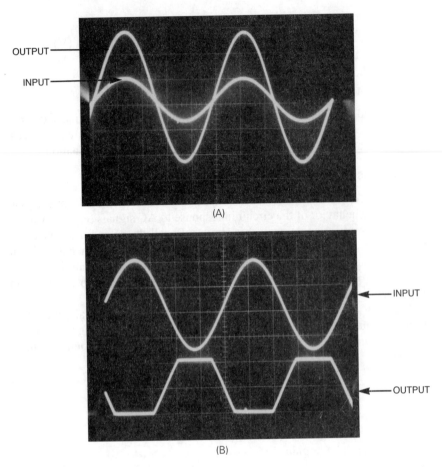

INPUT

(A)

INPUT

(B)

FIGURE 3-5 (A) Input and output waveforms for inverting amplifier working properly; (B) input and output waveforms with input signal too high (overdriving op-amp), causing clipping of output signal.

that sharp clipping occurs. Although there are better ways to realize this goal, the 'overdriven clipper' squarewave generator does work.

3-2.7 Response to AC input signals with DC offset

The case considered in the previous section assumed a waveform that is symmetrical about the zero volts baseline. In this section we will examine the case where an AC waveform is superimposed on a DC voltage. Figure 3-6 shows an inverting amplifier circuit with an AC signal source in series with a DC source. In Fig. 3-7A there is a 4 volt peak-to-peak squarewave superimposed on a 1 Vdc fixed potential. Thus, the non-symmetrical signal will swing between +3 volts and −1 volt.

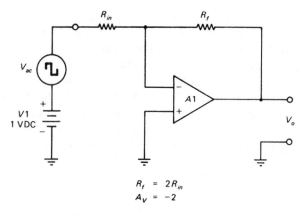

$$R_f = 2R_{in}$$
$$A_V = -2$$

FIGURE 3-6   Inverting amplifier with DC component to input signal.

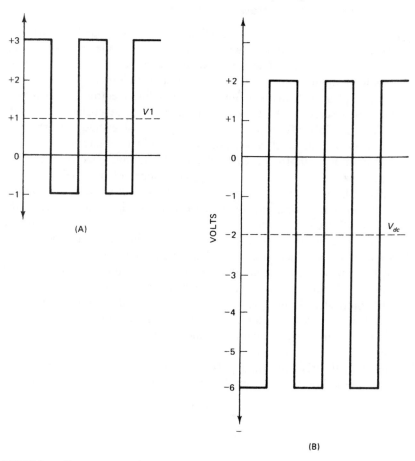

FIGURE 3-7   (A) Input waveform showing DC component; (B) resulting output waveform with DC offset.

The output waveform is shown in Fig. 3-7B. With the 180° phase inversion and the gain of −2 depicted in Fig. 3-6, the waveform will be a non-symmetrical oscillation between −6 volts and +2 volts. Because of gain ($A_v = -2$) the degree of asymmetry has also doubled to 2 Vdc.

Dealing with AC signals that have a DC component can lead to problems at high gain or high input signal levels. As was true in the case of the high amplitude symmetrical signal, the output may saturate at either $V-$ or $V+$ power supply rails. If this limit is reached, then clipping will result. The DC component is seen as a valid input signal so will drive the output to one power supply limit or the other. For example, if the op-amp in Fig. 3-6 has a $V_{o(max)}$ value of ±10 volts, then (with $A_v = -2$), a +4 volt positive input signal will saturate the output, while the negative excursion can reach −7 volts before causing output saturation.

3-2.8 Response to squarewaves and pulses

Most amplifiers respond in a congenial manner to sinusoidal and triangle waveforms. Some amplifiers, however, will exhibit problems dealing with fast risetime waveforms such as squarewaves and pulses. The source of these problems is the high frequency content of these waveforms.

All continuous mathematical functions (including electronic waveforms) are made of a series of harmonically related sine and cosine constituent waves (and possibly also a DC component). The sinewave consists of a single frequency sinusoidal wave. All non-sinusoidal waveforms, however, are made up of a single-frequency fundamental sinewave plus its harmonics. The actual waveshape is determined by the number of harmonics present, which particular harmonics are present (i.e. odd or even), the relative amplitudes of those harmonics, and their phase relationship with respect to the fundamental. These factors can be deduced from the quarter wave or halfwave symmetry of the wave. The listing of the constituent frequencies forms a *Fourier series*, and determines the bandwidth of the system required to process the signal. For example, the symmetrical squarewave is made up of a fundamental frequency sinewave (F), plus odd harmonics ($3F, 5F, 7F \cdots$) up to (theoretically) infinity (as a practical matter, most squarewaves are 'square' if the first 100 harmonics are present). Furthermore, if the squarewave is truly symmetrical, then all of the harmonics are in-phase with the fundamental. Other waveshapes have different Fourier spectrums.

In general, the risetime of a pulse is related to the highest significant frequency in the Fourier spectrum by the 'rule-of-thumb' approximation:

$$F = \frac{0.35}{T_r} \tag{3-33}$$

where:

F is the highest Fourier frequency in hertz (Hz)

T_r is the pulse risetime in seconds (s)

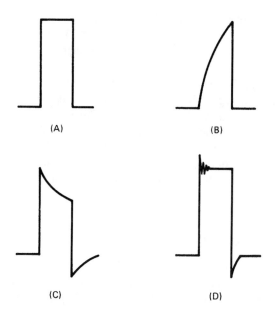

(A)

(B)

(C)

(D)

FIGURE 3-8 (A) Perfect squarewave; (B) squarewave with severe high frequency loss; (C) with severe low frequency loss; (D) with ringing.

Because pulse shape is a function of the Fourier spectrum for that wave, the frequency response characteristic of the amplifier has an effect on the waveshape of the reproduced signal. Figure 3-8 shows an input pulse signal (Fig. 3-8A) and two possible responses.

The response shown in Fig. 3-8B results from attenuation of the high frequencies. The rounding shown will be either moderate or severe depending on the −3 dB bandwidth of the amplifier. In other words, by how many harmonics are attenuated by the amplifier frequency response characteristic, and to what degree. This problem becomes especially severe when the fundamental frequency (or pulse repetition rate) is high, the risetime is very fast, and the amplifier bandwidth is low.

Frequency compensated operational amplifiers achieve their claimed 'unconditional stability' by rolling off the high frequency response drastically above a few kilohertz. A type 741AE is a frequency compensated op-amp with a gain–bandwidth (G–B) product of 1 250 000 Hz (i.e. 1.25 MHz) and an open-loop gain of 250 000. Frequency response at maximum gain is 1 250 000 Hz/250 000, or 5000 Hz. Thus, we can expect good squarewave response only at relatively low frequencies. A rule-of-thumb for squarewaves is to make the amplifier bandwidth at least 100 times the fundamental frequency. As with all such rules, however, this one should be applied with caution even though it is assimilated into your collection of 'standard engineering wisdom'.

The other class of problems is shown in Figs 3-8C and 3-8D. In this case we see *peaking* and *ringing* of the pulse. Three principal causes of these phenomena are found. First, a skewed bandpass characteristic in which either the low frequencies are attenuated (or amplified less) or the high frequencies are amplified more. Second, there are LC resonances in the circuit that give rise to ringing. Although not generally a problem at low frequencies, video operational amplifiers may see this problem. Third, there are both significant harmonics present at frequencies where circuit phase shifts add up to 180° and the loop gain is unity or greater. When combined with the 180° phase shift inherent in inverting followers, we have Barkhausen's criteria for oscillation. Under some conditions the device will break into sustained oscillation. In other cases, however, oscillation will occur only on fast risetime signal peaks as shown in Fig. 3-8D.

3-2.9 Some basic rules

We must consider several factors when designing inverting follower amplifiers. First, we obviously must consider the voltage gain required by the application. Second, we must consider the input impedance of the circuit. That specification is needed in order to prevent the amplifier input from loading the driving circuit. In the case of the inverting follower, the input impedance is the value of the input resistor (R_{in}), and a simple design rule is in effect:

The input resistor (hence the input impedance) should be equal or greater than ten times the source resistance of the previous circuit.[1]

The implication of this rule is that we must determine the source resistance of the driving circuit, and then make the input impedance of the operational amplifier inverting follower at least ten times larger. When the driving source is another operational amplifier we can assume that the source impedance (i.e. the output impedance of the driving op-amp) is 100 ohms or less. For these cases, make the value of R_{in} at least 1000 ohms (i.e. 10×100 ohms $=$ 1000 ohms). This value is based on consideration of available output load current. In other cases, however, we have a slightly different problem. Some transducers, for example a *thermistor* for measuring temperature, have a much higher source resistance. One thermistor has an advertised resistance that varies from 10 kohms to 100 kohms over the temperature range of interest, so a minimum input impedance of 1 megohm (i.e. 10×100 kohms) is required. When the input impedance becomes this high the designer might want to consider the noninverting follower rather than the inverting follower configuration.

---

[1] As with all 'rules of thumb', care must be exercised in its use. Many applications may require $100\times$ or $1000\times$.

In the inverting follower circuit the choice of input impedance drives the design, so is part of the design procedure:

Procedure

1. Determine the minimum allowable input resistance, i.e. ten or more times the source impedance).

2. If the source resistance is 1000 ohms or less, try 10 kohms as an initial trial input resistance (R_{in}). This value might be lowered if the feedback resistor (R_f) becomes too high for the required gain. The value of R_{in} is the input resistance, or 10 kohms whichever is higher.

3. Determine the amount of gain required. In general, the closed-loop gain of a single inverting follower should be less than 500. For gains higher than that figure use a multiple op-amp cascade circuit. Some low-cost op-amps should not be operated at closed-loop gains greater than 200. The reason for this rule is the problems that are found in real (versus ideal) devices (see Chapter 4). In those cases the distributed gain of a cascade amplifier may prove easier to tame in practical situations.

4. Determine the frequency response (i.e. the frequency at which the gain drops to unity). From steps 3 and 4 we can calculate the minimum gain bandwidth product of the op-amp required (G–B = gain × frequency).

5. Select the operational amplifier. If the gain is high, e.g. over 100, then you might want to select a BiMOS or BiFET operational amplifier in order to limit the output offset voltage caused by the input bias currents. Select a 741-family device if (a) you don't need more than a few kilohertz frequency response, and (b) the unconditionally stable characteristics of the 741 is valuable for the application. Also look at the package style. For most applications the 8-pin miniDIP package is probably the easiest to handle. The 8-pin metal can is also useful, and it can be made to fit 8-pin miniDIP positions by correct bending of the leads.

6. Select the value of the feedback resistor:

$$R_f = ABS(A_v) \times R_{in} \tag{3-34}$$

(Note: ABS(A_v) means the absolute value of voltage gain, i.e. without the minus sign.)

7. If the value of the feedback resistor is too high, i.e. beyond the range of standard values (about 20 megohms or so), or, too high for the input bias currents, then try a lower input resistance.

3-2.10 Altering AC frequency response

The natural bandwidth of an amplifier is sometimes too great for certain specific applications. Noise power, for example, is a function of bandwidth as indicated by the expression $P_n = KTBR$. Thus, it is possible that signal-to-noise ratio will suffer in some applications if bandwidth is not limited to that which is actually needed to process the expected waveforms. In other cases we find that the rejection of spurious signals suffers if we fail to tailor the bandwidth of an amplifier circuit to that which is required by the bandwidth of the applied input signal. Amplifier stability is improved if the loop gain of the circuit is reduced to less than the frequency at which the circuit phase shifts (including internal amplifier phase shift) reaches 180°. When the distributed phase shift is added to the 180° phase shift seen normally on inverting amplifiers, Barkhausen's criteria (360° feedback) is satisfied, and the amplifier will oscillate. Those criteria are:

1. Total phase shift of 360° at the frequency of oscillation;

2. Output-to-input coupling (may be accidental); and

3. Loop gain of unity or greater.

If these criteria are satisfied at any frequency, then the operational amplifier will oscillate at that frequency. We will discuss this topic further in Section 3-4, in Chapter 4 when we discuss operational amplifier problems. For the present we will discuss just one technique in case you need to know the method in performing laboratory exercises.

The design goal in tailoring the AC frequency response is to roll-off the voltage gain at the frequencies above a certain critical frequency, F_c. This frequency is determined by evaluating the application, and is defined as the frequency at which the gain of the circuit drops off -3 dB from its in-band voltage gain. The response of the amplifier should look like Fig. 3-9A, and is shown in here in normalized form in which the maximum in-band gain is taken to be 0 dB. Above the critical frequency the gain drops off -6 dB/octave (an octave is a 2:1 change in frequency) by shunting a capacitor across the feedback resistor, as shown in Fig. 3-9B. The reactance of the capacitor is shunted across the resistance of R_f, so reduces the gain. the low-pass filter characteristic is achieved because the capacitive reactance becomes lower as frequency increases. The value of the capacitor is found from:

$$C = \frac{1}{2\pi R_f F_c} \qquad (3\text{-}35)$$

where:
   C is the capacitance in farads (F)
   R_f is the feedback resistance in ohms (Ω)
   F_c is the -3 dB frequency in hertz (Hz)

Alternatively, to calculate the capacitance of C in microfarads (μF) we use Eq. (3-36):

$$C_{\mu F} = \frac{1\,000\,000}{2\pi R_f F_c} \qquad (3\text{-}36)$$

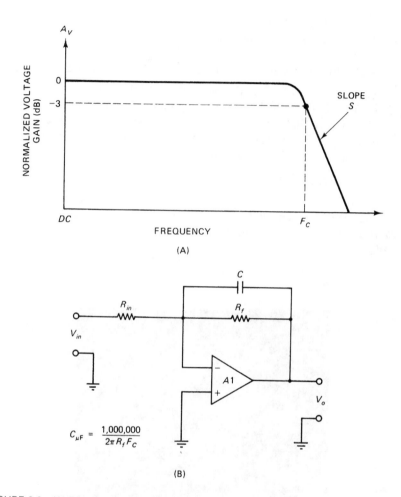

FIGURE 3-9 (A) Frequency response limiting as measured from the −3 dB point; (B) using a capacitor to frequency limit the response of an inverting follower.

EXAMPLE 3-5

An inverting follower with a gain of −100 has a 470 kohm feedback resistor. Calculate the capacitance (in μF) if the −3 dB point in the response curve is 200 Hz.

Solution

$$C2_{\mu F} = \frac{1\,000\,000}{2\pi R_f F_c} \qquad \text{(Eq. (3-36) restated)}$$

$$C2_{\mu F} = \frac{1\,000\,000}{(2)(3.14)(470\,000 \text{ k}\Omega)(200 \text{ Hz})}$$

$$C2_{\mu F} = \frac{1\,000\,000}{5.9 \times 10^8} = 0.00169 \text{ μF} \qquad ■$$

In the example above a standard value of either 0.0015 μF or 0.002 μF would probably be used as a practical matter.

3-3 NONINVERTING FOLLOWERS

The next standard op-amp circuit configuration is the *noninverting follower*. This type of amplifier uses the noninverting input of the operational amplifier to apply signal. In this configuration the output signal is in-phase with the input signal (Fig. 3-10). There are two basic noninverting configurations: *unity gain* and *greater-than-unity gain*.

3-3.1 Unity gain follower

Figure 3-11 shows the circuit for the unity gain noninverting follower. The output terminal is connected directly to the inverting input, resulting in 100% negative feedback. Recall the voltage gain expression for all feedback amplifiers:

$$A_v = \frac{A_{vol}C}{1 + A_{vol}\beta} \tag{3-37}$$

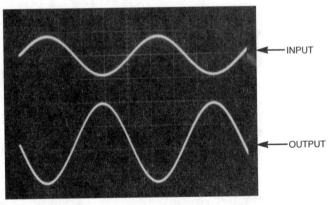

FIGURE 3-10   Input and output waveforms superimposed.

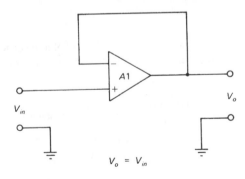

FIGURE 3-11   Unity gain noninverting follower amplifier.

where:

A_v is the closed-loop voltage gain (i.e. gain with feedback)

A_{vol} is the open-loop voltage gain (i.e. gain without feedback)

β is the feedback factor

C is the input attenuation factor

In this circuit the input signal is applied directly to +IN, so $C = 1$, and can therefore be ignored. The feedback factor, β, represents the transfer function of the feedback network. When that network is a resistor voltage divider network, the value of β is a decimal fraction that represents the attenuation of the op-amp output voltage before it is applied to the op-amp inverting input. In the unity gain follower circuit the value of β is also 1 so it too is ignored. Equation (3-24) therefore reduces to:

$$A_v = \frac{A_{vol}}{1 + A_{vol}\beta} \qquad (3\text{-}38)$$

Consider the implications of Eq. (3-37) for a common operational amplifier:

EXAMPLE 3-6

A type 741 device has an open-loop gain of 300 000. For this device, the voltage gain of a circuit such as Fig. 3-11 is:

Solution

$$A_v = \frac{A_{vol}}{1 + A_{vol}\beta} \quad \text{(Eq. (3-38) restated)}$$

$$A_v = \frac{300\,000}{1 + 300\,000}$$

$$A_v = \frac{300\,000}{300\,001} = 0.9999967 \qquad \blacksquare$$

A gain of 0.9999967 is close enough to 1.0 to justify calling the circuit of Fig. 3-11 a unity gain follower.

3-3.2 Applications of unity gain followers

What use is an amplifier that does not amplify? There are three principal uses of the unity gain noninverting follower: *buffering, power amplification,* and *impedance transformation*.

A 'buffer' amplifier is placed between a circuit and its load in order to improve the isolation between the two. An example is use of a buffer amplifier between an oscillator or waveform generator and its load. The buffer is especially useful where the load exhibits a varying impedance that could result in 'pulling' of the oscillator frequency. Such unintentional frequency modulation of the oscillator is very annoying because it makes some oscillator circuits unable to function and causes others to function poorly.

Another common use for buffer amplifiers is isolation of an output connection from the main circuitry of an instrument. An example might be an instrumentation circuit that uses multiple outputs, perhaps one to a digital computer A/D converter input and another to an analog oscilloscope or strip chart recorder. By buffering the analog output to the oscilloscope we prevent short circuits in the display wiring from affecting the signal to the computer, and vice versa.

A special case of buffering is represented by using the unity gain follower as a power driver. A long cable run may attenuate low-power signals. To overcome this problem we sometimes use a low impedance power source to drive a long cable. This application points out that a unity gain follower does have power gain (the unity gain feature refers only to the voltage gain). If the input impedance is typically much higher than the output impedance, yet $V_o = V_{in}$, then by V^2/R it stands to reason that the delivered power output is much greater than the input power. Thus, the circuit of Fig. 3-11 is unity gain for voltage signals and greater than unity gain for power. It is therefore a power amplifier.

The impedance transformation capability is obtained from the fact that an op-amp has a very high input impedance and a very low output impedance. Let's illustrate this application by a practical example. Figure 3-12A is a generic equivalent of a voltage source driving a load ($R2$). The resistance $R1$ represents the internal impedance of the signal source (usually called 'source impedance'). The signal voltage, V, is reduced at the output (V_o) by whatever voltage is dropped across source resistance $R1$. The output voltage is found from:

$$V_o = \frac{V\ R2}{R1 + R2} \tag{3-39}$$

EXAMPLE 3-7

Assume that a high impedance source is driving a low impedance load similar to Fig. 3-12A. Calculate the loss in a case where $R1 = 10$ kohms, $R2 = 1$ kohm, and $V = 1$ volt.

Solution

$$V_o = \frac{V\ R2}{R1 + R2} \quad \text{(Eq. (3-39) restated)}$$

$$V_o = \frac{(1\ \text{V})(1\ \text{k}\Omega)}{10\ \text{k}\Omega + 1\ \text{k}\Omega}$$

$$V_o = \frac{1}{11}\ \text{volts} = 0.091\ \text{volts} \qquad \blacksquare$$

In Example 3-7, 90% of the signal voltage was lost by the voltage divider action. But if we impedance transform the circuit with a unity gain noninverting amplifier, as in Fig. 3-12B, then we change the situation entirely.

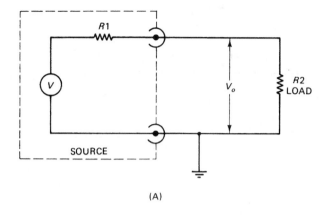

(A)

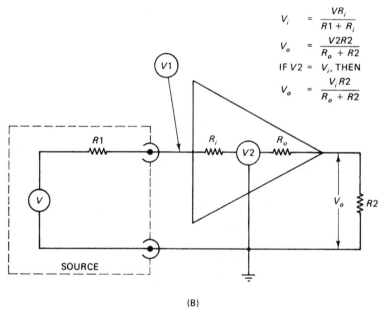

$$V_i = \frac{VR_i}{R1 + R_i}$$

$$V_o = \frac{V2R2}{R_o + R2}$$

IF $V2 = V_i$, THEN

$$V_o = \frac{V_i R2}{R_o + R2}$$

(B)

FIGURE 3-12 (A) Equivalent circuit showing effect of internal resistance ($R1$) on output voltage; (B) equivalent circuit when source is in cascade with an amplifier input.

If the amplifier input impedance is very much larger than the source resistance, and the amplifier output impedance is very much lower than the load impedance, then there is very little loss and V will closely approximate V_o.

3-3.3 Noninverting followers with gain

Figure 3-13A shows the circuit for the noninverting follower with gain. In this circuit, the signal (V_{in}) is applied to the noninverting input, while the

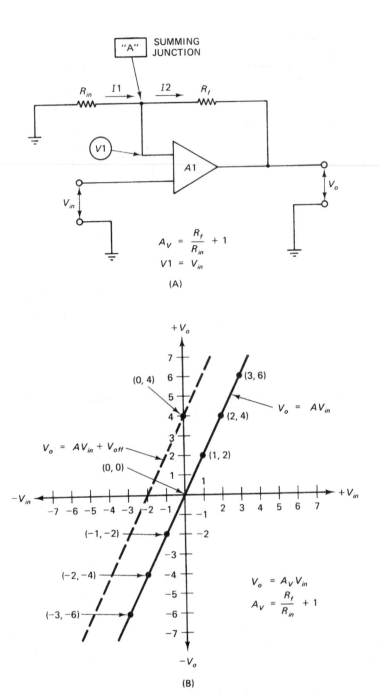

FIGURE 3-13 (A) Circuit for noninverting follower with gain; (B) transfer function for (A).

feedback network (R_f/R_{in}) is almost the same as it was in the inverting follower circuit. The difference is that one end of R_{in} is grounded.

We can evaluate this circuit using the same general method as was used in the inverting follower case. We know from Kirchhoff's current law, and the fact that the op-amp inputs neither sink nor source current, that I_1 and I_2 are equal to each other. Thus, the Kirchhoff expression for these currents at the summing junction (point A) can be written as:

$$I_1 = I_2 \qquad (3\text{-}40)$$

We know from the properties of the ideal op-amp that any voltage applied to the noninverting input (V_{in}) also appears at the inverting input. Therefore:

$$V_1 = V_{in} \qquad (3\text{-}41)$$

From Ohm's law we know that the value of current I_1 is:

$$I_1 = \frac{V_1}{R_f} \qquad (3\text{-}42)$$

or, because $V_1 = V_{in}$,

$$I_1 = \frac{V_{in}}{R_f} \qquad (3\text{-}43)$$

Similarly, current I_2 is equal to the voltage drop across resistor R_f divided by the resistance of R_f. The voltage drop across resistor R_f is the difference between output voltage V_o and the voltage found at the inverting input, V_1. By ideal property No. 7, $V_1 = V_{in}$; therefore:

$$I_2 = \frac{V_o - V_{in}}{R_f} \qquad (3\text{-}44)$$

We can derive the transfer equation of the noninverting follower by substituting Eqs (3-42) and (3-43) into Eq. (3-39):

$$\frac{V_{in}}{R_{in}} = \frac{V_o - V_{in}}{R_f} \qquad (3\text{-}45)$$

We must now solve Eq. (3-44) for output voltage V_o:

$$\frac{V_{in}}{R_{in}} = \frac{V_o - V_{in}}{R_f} \qquad (3\text{-}46)$$

$$\frac{R_f - V_{in}}{R_{in}} = V_o - V_{in} \qquad (3\text{-}47)$$

$$\frac{R_f - V_{in}}{R_{in}} + V_{in} = V_o \qquad (3\text{-}48)$$

Factoring out V_{in}:

$$V_{in} \times \left[\frac{R_f}{R_{in}} + 1 \right] = V_o \qquad (3\text{-}49)$$

or, reversing the order to the conventional style:

$$V_o = V_{in} \left[\frac{R_f}{R_{in}} + 1 \right] \qquad (3\text{-}50)$$

Equation (3-50) is the transfer equation for the noninverting follower.

The transfer function V_o/V_{in} for a gain of -2 noninverting amplifier is shown in Fig. 3-13B. The solid line assumes no output offset voltage is present (i.e. $V_o = A_v V_{in} + 0$), while the dotted line represents a case where the offset voltage is non-zero.

EXAMPLE 3-8

A noninverting amplifier has a feedback resistor of 100 kohms, and an input resistor of 20 kohms. What is the voltage gain?

Solution

$$A_v = \frac{R_f}{R_{in}} + 1$$

$$a_v = \frac{100 \text{ k}\Omega}{20 \text{ k}\Omega} + 1 = 5 + 1 = 6 \qquad \blacksquare$$

3-3.4 Advantages of noninverting followers

The noninverting follower offers several advantages. In our discussion of the unity gain configuration we mentioned that buffering, power amplification and impedance transformation were advantages. Also, in the gain noninverting amplifier configuration we are able to provide voltage gain with no phase reversal.

The input impedance of the noninverting followers shown thus far is very high, being essentially the input impedance of the op-amp itself. In the ideal device, this impedance is infinite, while in practical devices it may range from 500 000 to more than 10^{12} ohms. Thus, the noninverting follower is useful for amplifying signals from any high impedance source, regardless of whether or not impedance transformation is a circuit requirement.

When the gain required is known (as it usually is in practical situations) we select a trial value for R_{in}, and then solve Eq. (3-50) to find R_f. This new version of the equation is:

$$R_f = R_{in}(A_v - 1) \qquad (3\text{-}51)$$

We can solve many design problems using Eq. (3-51).

EXAMPLE 3-9

Design a gain-of-100 noninverting amplifier.

Solution

1. Select a trial value of resistance for R_{in} (typically 100 to 5000 ohms): $R_{in} =$ 1200 ohms.

2. Use Eq. (3-51) to determine the value of R_f:

$$R_f = R_{in}(A_v - 1)$$

$$R_f = (1200\ \Omega)(100 - 1) = (1200\ \Omega)(99) = 118\,800\ \Omega \qquad \blacksquare$$

Determine by evaluating the application whether or not the trial result obtained from this operation is acceptable. If the result is not acceptable, then work the problem again using a new trial value.

What does 'acceptable' mean? If the value of R_f is exactly equal to a standard resistor value, then all is well. But, as in the case above, the value (118 800 ohms) is not a standard value. What we have to determine, therefore, is whether or not the nearest standard values result in an acceptable gain error (which is determined from the application). Both 118 kohms and 120 kohms are standard values, with 120 kohms being somewhat easier to obtain for distributor stock inventories. Both of these standard values are less than 1% from the calculated value, so this result is acceptable if a 1% gain error is within reasonable tolerance limits for the application.

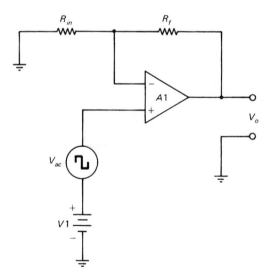

FIGURE 3-14　Noninverting amplifier with DC component in series with the signal.

3-3.5 The AC response of noninverting amplifiers

The noninverting amplifier circuits discussed in the preceding sections are DC amplifiers. Nonetheless, as with the inverting amplifiers considered earlier, the noninverting amplifier will also respond to AC signals up to the upper

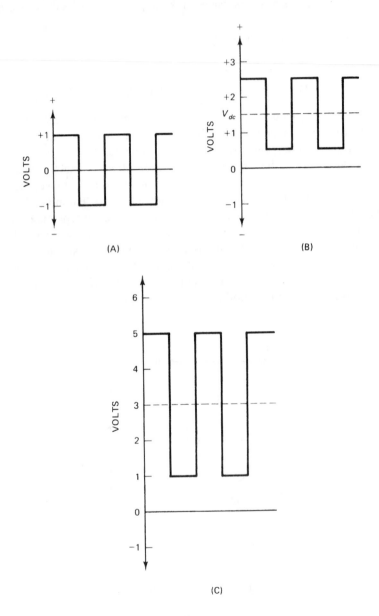

(A)

(B)

(C)

FIGURE 3-15   Effects of increasing DC offset.

frequency response limit of the circuit. It is recommended that you go back to Sections 3-2.6 and 3-2.7 for a review.

Figure 3-14 shows the input signal situation for a noninverting follower. In this case there is an AC signal source in series with a DC potential ($V1$), which are applied to the noninverting input of the operational amplifier. A squarewave input signal (Fig. 3-15A) is applied to the input, but it is offset by a DC component (Fig. 3-15B). If the amplifier has a gain of +2, then the output signal will be as shown in Fig. 3-15C. This signal swings from +1 volt to +5 volts. The offset of 1.5 Vdc is amplified by two, and becomes a 3 Vdc offset, with the AC signal swinging about this level.

3-4 FREQUENCY RESPONSE TAILORING

In Section 3-2.10 we learned that we can tailor the upper-end frequency response of the inverting follower operational amplifier with a capacitor shunting the feedback resistor. The same method also works for the noninverting follower. In this section we will expand the subject and discuss not only tailoring of the upper −3 dB frequency response, but also a lower −3 dB limit as well.

The capacitor across the feedback resistor (Fig. 3-16) sets the frequency at which the upper end frequency response falls off −3 dB below the low end in-band gain. The gain at frequencies higher than this −3 dB frequency falls off at a rate of −6 dB/octave (2:1 frequency ratio), or −20 dB/decade (10:1 frequency ratio).

The value of capacitor $C2_{\mu F}$ is found by:

$$C2 = \frac{1\,000\,000}{2\pi R_f F_c} \tag{3-52}$$

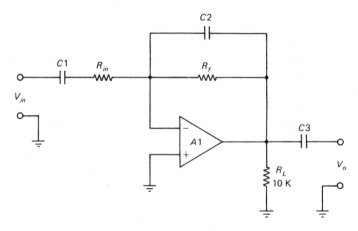

FIGURE 3-16   AC-coupled inverting follower amplifier.

where:

$C2_{\mu F}$ is the capacitance in microfarads (μF)

R_f is in ohms (Ω)

F_c is the upper -3 dB frequency in hertz (Hz)

The low frequency response is controlled by placing a capacitor in series with the input resistor, which makes the inverting follower an AC-coupled amplifier. Figure 3-16 is the circuit for an inverting follower that uses AC-coupling at both input and output circuits. Capacitor $C2$ limits the upper -3 dB frequency response point. Its value is set by the method discussed above. The lower -3 dB point is set by the combination of R_{in} and input capacitor $C1$. This frequency is set by the equation:

$$C1_{\mu F} = \frac{1\,000\,000}{2\pi R_{in} F} \tag{3-53}$$

where:

$C1_{\mu F}$ is in microfarads (μF)

R_{in} is in ohms (Ω)

F is the lower -3 dB point in hertz (Hz)

In some cases we will want to AC-couple the output circuit (although it is optional in most cases). Capacitor $C3$ is used to AC-couple the output, thus preventing any DC component that is present on the op-amp output from affecting the following stages. Resistor R_L is used to keep capacitor $C3$ from being charged by the offset voltage from op-amp $A1$. The value of capacitor $C3$ is set to retain the lower -3 dB point, using the resistance of the stage following as the 'R' in the equations above.

3-5 AC-COUPLED NONINVERTING AMPLIFIERS _____

The noninverting amplifiers discussed thus far have all been DC-coupled. They will respond to signals from either DC or near-DC up to the frequency limit of the amplifier selected. Sometimes, however, we do not want the amplifier to respond to DC or slowly varying near-DC signals. For these applications we select an AC-coupled noninverting follower circuit. In this section we will examine several AC-coupled noninverting amplifiers.

Figure 3-17A shows a capacitor input AC-coupled amplifier circuit. It is essentially the same as the previous circuits, except for the input coupling network, $C1/R3$. The capacitor in Figure 3-17A serves to block DC and very low frequency AC signals. If the op-amp has zero (or, more realistically, very low) input bias currents, then we can safely delete resistor $R3$. For all but a few commercially available devices, however, resistor $R3$ is required if closed-loop gain is high. Input bias currents will charge capacitor $C1$, creating a voltage offset that is seen by the op-amp as a valid DC signal and amplified

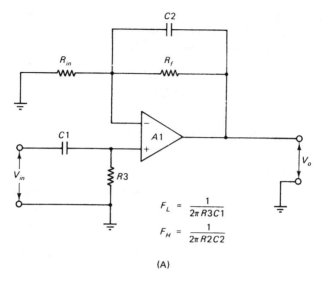

$$F_L = \frac{1}{2\pi R3C1}$$

$$F_H = \frac{1}{2\pi R2C2}$$

(A)

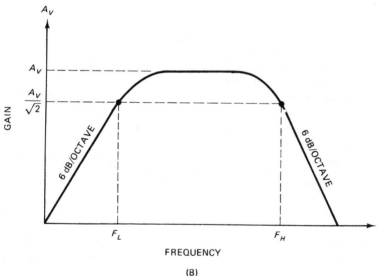

(B)

FIGURE 3-17   (A) AC-coupled noninverting follower amplifier; (B) typical frequency response curve.

to form an output offset voltage. In some devices the output saturates from the $C1$ charge shortly after turn-on; resistor $R3$ keeps $C1$ discharged.

Resistor $R3$ also sets the input impedance of the amplifier. Previous circuits had a very high input impedance because that parameter was determined only by the (extremely high) op-amp input impedance. In Fig. 3-17A, however, the input impedance seen by the source is equal to $R3$.

Another effect of resistor $R3$ and capacitor $C1$ is to limit the low frequency response of the circuit. Filtering occurs because $R3C1$ forms a high-pass filter (see Fig. 3-17B). The -3 dB frequency is found from:

$$F = \frac{1\,000\,000}{2\pi R_{in} F} \tag{3-54}$$

where:
   F is the -3 dB frequency in hertz (Hz)
   $R1$ is the resistance in ohms (Ω)
   $C1$ is the capacitance in microfarads (μF)

The form of Eq. (3-54) is backwards from the point of view of practical circuit design problems. In most cases, we know the required frequency response limit from the application. We also know from the application what the minimum value of $R3$ should be (an implication from source impedance), and often set it as high as possible as a practical matter (e.g. 10 megohms). Thus, we want to solve the equation for C, as shown below:

$$C_{\mu F} = \frac{1\,000\,000}{2\pi R_1 F} \tag{3-55}$$

(all terms as defined above).

The technique of Fig. 3-17A works well for dual-polarity DC power supply circuits. In single-polarity DC power supply circuits, however, the method falls down because of the large DC offset voltage present on the output. For these applications we use a circuit such as shown in Fig. 3-18.

The circuit in Fig. 3-18 is operated from a single $V+$ DC power supply (the $V-$ terminal of the op-amp is grounded). In order to compensate for the $V-$ supply being grounded, the noninverting input is biased to a potential of:

$$V1 = (V+)\frac{R4}{R4 + R5} \tag{3-56}$$

If $R4 = R5$, then $V1$ will be $(V+)/2$. Because the noninverting input typically sinks very little current, the voltage at both ends of $R3$ is the same (i.e. $V1$).

The circuit of Fig. 3-18 does not pass DC and some low AC frequencies because of the capacitor coupling. Also, because capacitor $C3$ shunts feedback resistor $R2$, there is also a roll-off of the higher frequencies. The high frequency roll-off -3 dB point is found from:

$$F_{Hz} = \frac{1\,000\,000}{2\pi R2 C3_{\mu F}} \tag{3-57}$$

where:
   F is the -3 dB frequency in hertz (Hz)

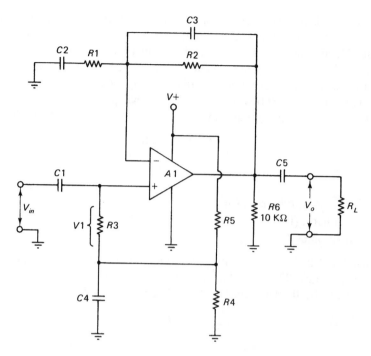

FIGURE 3-18 Single-supply operation of the AC-coupled operational amplifier.

$R2$ is in ohms (Ω)

$C3_{\mu F}$ is in microfarads (μF)

We can restate Eq. (3-57) in a more useful form that takes into account the fact that we usually know the value of $R2$ (from setting the gain), and the nature of the application sets the minimum value of frequency, F. We can rewrite Eq. (3-56) in a form that yields the value of $C3$ from these data:

$$C3_{\mu F} = \frac{1\,000\,000}{2\pi R3 F_{Hz}} \qquad (3\text{-}58)$$

The lower -3 dB frequency is set by any or all of several RC combinations within the circuit:

1. $R1$ and $C2$
2. $R3$ and $C1$
3. $R3$ and $C4$
4. R_L and $C5$

Resistor $R1$ is part of the gain-setting feedback network. Capacitor $C2$ is used to keep the 'cold' end of $R1$ above ground at DC, while keeping it grounded for AC signals.

Resistor $R3$ is the input resistor and serves the same purpose as the similar resistor in the previous circuit. At midband the input impedance is set by resistor $R3$, although at the extreme low end of the frequency range the reactance of $C4$ becomes a significant feature. In general, X_{C4} should be less than or equal to $R3/10$ at the lowest frequency of operation.

Capacitor $C1$ is in series with the input signal path and serves to block DC and certain very low frequency signals. The value of $C1$ should be:

$$C1_{\mu F} = \frac{1\,000\,000}{2\pi F R_L} \qquad (3\text{-}59)$$

where:
   $C1\mu F$ is in microfarads (μF)
   F is in hertz (Hz)
   $R3$ is in ohms (Ω)

Capacitor $C5$ is used to keep the DC output offset from affecting succeeding stages. The 10 kohm output load resistor ($R6$) keeps $C5$ from being charged by the DC offset voltage. The value of $C5$ is given by:

$$C5 \geq \frac{1\,000\,000}{2\pi F R_L} \qquad (3\text{-}60)$$

where:
   $C5_{\mu F}$ is in microfarads (μF)
   F is the low end -3 dB frequency in hertz (Hz)
   R_L is the load resistance in ohms (Ω)

3-6 TRANSFORMER-COUPLED NONINVERTING AMPLIFIERS

Figure 3-19 shows the circuit for a transformer-coupled noninverting follower. This type of circuit is often used in audio and broadcasting applications. In those applications we might find audio signals passing over a 600 ohm balanced line. The point we will make here is that this circuit is an AC-only amplifier, with upper and lower -3 dB points determined by (a) the frequency response of the transformer ($T1$), the limitations of the operational amplifier, and any capacitances shunting feedback resistor $R2$.

The gain of the amplifier in Fig. 3-19 is given in Eq. (3-61) below:

$$A_v = V_{in} \left[\frac{N_s}{N_p} \right] \left[\frac{R2}{R1} + 1 \right] \qquad (3\text{-}61)$$

where:
   A_v is the voltage gain
   N_s is the number of turns in the secondary of $T1$

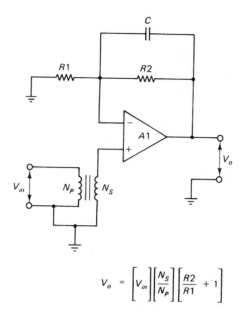

$$V_o = \left[V_{in} \right]\left[\frac{N_S}{N_P} \right]\left[\frac{R2}{R1} + 1 \right]$$

FIGURE 3-19   Transformer-coupled input to noninverting follower.

N_p is the number of turns in the primary of $T1$
$R2$ is the op-amp feedback resistor
$R1$ is the op-amp input resistor

(Note: $R1$ and $R2$ in same units: ohms, kohms or megohms.)

EXAMPLE 3-10

Find the gain of an amplifier such as Fig. 3-19 if transformer $T1$ has a turns ratio (N_p/N_s) of 1:4.5 (which means that $N_s/N_p = 4.5 : 1$), $R1 = 1000$ ohms and $R2 = 3900$ ohms.

Solution

$$A_v = V_{in} \left[\frac{N_s}{N_p} \right] \left[\frac{R2}{R1} + 1 \right] \quad \text{(Eq. (3-61) restated)}$$

$$A_v = \left[\frac{4.5}{1} \right] \left[\frac{3900 \ \Omega}{1000 \ \Omega} + 1 \right]$$

$$A_v = (4.5)(3.9 + 1) = (4.5)(4.9) = 22.1 \qquad \blacksquare$$

3-7   SUMMARY

1. There are three principal configurations for the operational amplifier: inverting follower, unity gain noninverting follower, and noninverting follower with gain.

2. The inverting amplifier produces an output signal that is 180° out of phase with the input signal.

3. A noninverting amplifier produces a signal that is in-phase with the input signal.

4. A virtual ground is any point where the potential is zero volts, i.e. ground potential, even though the point is not physically connected to ground or the power supply common.

5. The gain of an operational amplifier circuit is set by the external feedback resistor network.

3-8 RECAPITULATION

Now return to the objectives and Pre-quiz questions at the beginning of the chapter and see how well you can answer them. If you cannot answer certain questions, place a check mark to each and review the appropriate parts of the text. Next, try to answer the questions and work the problems below, using the same procedure.

3-9 LABORATORY EXERCISES

1. Connect the circuit of Fig. 3-20 using values of R_f selected from the range 22 kohms to 56 kohms (try several). With each value of feedback resistor make a graph of V_o versus V_{in} for five values of $+V_{in}$ and five values of $-V_{in}$.

2. Using the circuit of Fig. 3-20, and a value of R_f of 150 kohms, make a plot of V_o versus V_{in} similar to that required in the previous exercise. Explain the difference between the two plots.

3. Replace the DC voltage source in Fig. 3-20 with a 100 Hz sinewave source. Observe the input and output waveforms on a cathode ray oscilloscope (CRO).

4. Design a noninverting amplifier with a gain of: (a) 2; and (b) 101.

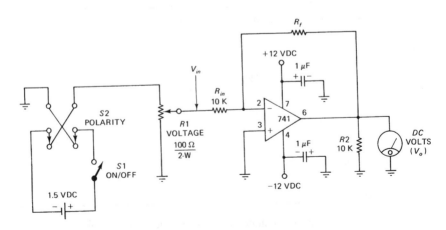

FIGURE 3-20   Test circuit.

3-10 QUESTIONS AND PROBLEMS _____

1. The potential measured at a properly functioning virtual ground is: _____ −Vdc.

2. An operational amplifier is connected as an inverting follower. The feedback resistor is 100 kohms, and the input resistor is 15 kohms. (a) What is the gain of this circuit? (b) What is the output voltage if a 0.800 Vdc input signal voltage is applied?

3. Consider the circuit of Fig. 3-21. Sketch one cycle of the output voltage, V_o if the input signal is 1 volt (peak-to-peak) at 100 Hz, and $V1$ is 1.5 Vdc. [Hint: $V1$ and $V2$ see different operational amplifier configurations].

4. In the circuit of Fig. 3-21 point A is grounded, and a +2.5 Vdc potential is applied to the noninverting input ($V1$). (a) Which amplifier configuration is this new circuit? (b) What is the output voltage? (c) What voltage is measured at point B?

5. Consider Fig. 3-22. Calculate V_o for the following conditions: $V1 = 1$ Vdc, $V2 = 0.4$ Vdc, $V3 = 2$ Vdc, and $V4 = −1.5$ Vdc.

6. Calculate V_o in Fig. 3-22 if $V1 = 2$ Vdc, $V2 = 1$ Vdc, $V3 = 560$ mV, and $V4 = 2.75$ Vdc.

7. A 741 operational amplifier is connected to ±12 Vdc power supplies. In this amplifier the maximum value of V_o is 1 volt less than the DC power supply potential of the same polarity. If $R_f = 100$ kohm, and $R_{in} = 10$ kohms, will this amplifier saturate with input signals of (a) 100 mV, (b) 1 Vdc, (c) 1 volt p-p 100 Hz AC, (d) 5 Vdc, (e) −5 Vdc.

8. What is the maximum allowable input signal voltage if the closed-loop gain of an inverting follower is 150, and the maximum allowable output voltage is 14 volts.

9. Sketch the output waveform from a gain of +2 noninverting follower if the input waveform is a 2 volt peak-to-peak squarewave with a 1 Vdc offset.

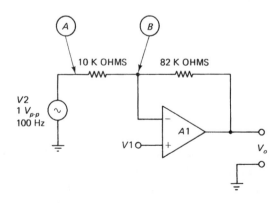

FIGURE 3-21

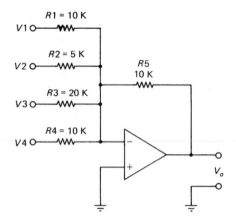

FIGURE 3-22

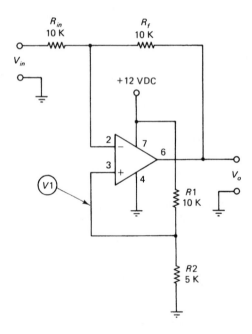

FIGURE 3-23

10. Calculate the approximate highest Fourier frequency in a squarewave with a 1.6 μs risetime.

11. State Barkhausen's criteria for oscillation, and explain how they apply to self-oscillation of an operational amplifier.

12. A signal source has an internal impedance of 15 kohms. What is the minimum rule-of-thumb input impedance of an inverting follower that is used to amplify the output signal of this signal source?

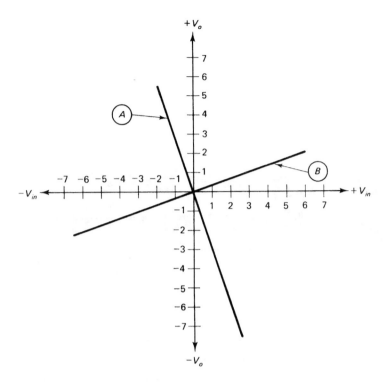

FIGURE 3-24

13. A noninverting follower has a 470 kohm feedback resistor. Calculate the value of shunt capacitor needed to limit the -3 dB frequency response to 100 Hz.

14. List three uses for a unity gain noninverting follower.

15. Consider Fig. 3-23. (a) Calculate $V1$; (b) calculate V_o when V_{in} is 0 Vdc; (c) sketch the output waveform when V_{in} is a 100 Hz, 5 volt peak-to-peak sinewave.

16. Consider the V_o versus V_{in} characteristic plotted in Fig. 3-24. (a) What is the gain of the amplifier depicted as curve A. (b) What is the gain of the amplifier depicted as curve B. (c) Design amplifiers that will exhibit these characteristics.

17. Using Fig. 3-24 as a guide draw characteristics for the following gains: (a) $+5$; (b) -5; (c) $+10$; (d) -1; and (e) -6.

CHAPTER 4

Dealing with practical operational amplifiers

OBJECTIVES

1. Understand problems existing in real versus ideal operational amplifiers.
2. Understand how to deal with offset voltages in real operational amplifiers.
3. Learn how to frequency compensate operational amplifiers.
4. Know how to measure the parameters of operational amplifiers.

4-1 PRE-QUIZ

These questions test your prior knowledge of the material in this chapter. Try answering them before you read the chapter. Look for the answers (especially those you answered incorrectly) as you read the text. After you have finished studying the chapter try answering these questions again, and those at the end of the chapter (see Section 4-9).

1. An input bias offset current of 3.7 mA flows in the inputs of a 741 operational amplifier. Calculate the input offset voltage on the inverting input if R_f is 100 kohms, and R_{in} is 10 kohms.

2. What are the implications of not decoupling the $V-$ and $V+$ power supply lines of an uncompensated IC linear amplifier?

3. Why are two different forms of capacitor sometimes used in power supply decoupling of IC linear amplifiers?

4. The unity gain capacitance for an operational amplifier is 68 pF. Find the value of a compensation capacitor if R_f is 100 kohms, and R_{in} is 2.7 kohms.

4-2 OPERATIONAL AMPLIFIER PROBLEMS AND THEIR SOLUTION

In Chapter 2 the ideal operational amplifier was discussed. Such a hypothetical device is a good learning tool, but doesn't really exist. While it makes our analysis easier, it cannot actually be purchased and used in practical circuits. All real operational amplifiers depart somewhat — in some areas a great deal — from the ideal. We find, for example, that open-loop gain is not really infinite, but rather very high values in the range from about 20 000 to over 1 000 000 are found. Similarly, real operational amplifiers don't have infinite bandwidth, and in fact many types are intentionally made severely bandwidth limited. Such amplifiers are said to be *unconditionally stable* or *frequency compensated* devices (an example is the 741 device). While stability is a highly desirable feature for many applications, it is obtained at the expense of frequency response. In this chapter we will deal with some of the more common problems found in real devices, and their solutions. We will also take a look at how various amplifier parameters are measured.

4-3 MEASUREMENT OF OPERATIONAL AMPLIFIER PARAMETERS

In this section we will look at how certain amplifier parameters are measured. The purpose of this discussion is two-fold. First, the procedure will allow you to make your own measurements in the laboratory on real devices. Second, and perhaps most important, is that you will understand specification parameters and how they are derived. Many op-amp parameters can only be understood in the context of the measurement situation. Comparison of devices becomes nearly impossible if different test methods are used for each device. For example, thermal drift can be compared in two devices only if both are measured at the same $V-$ and $V+$ power supply potentials.

4-3.1 Voltage gain

Voltage gain is the ratio of signal output voltage (V_o) to signal input voltage (V_{in}), or in equation form:

$$A_v = \frac{V_o}{V_{in}} \tag{4-1}$$

Measuring the closed-loop voltage gain is a simple matter of applying an input signal of known amplitude, and then measuring the output level generated. The voltage gain is calculated from Eq. (4-1). The process is simple, but watch for saturation of the output signal. An AC input signal such as a sinewave or triangle wave is preferred for this measurement because 'flat-topping' (the indicator of saturation) is easier to spot.

The input signal must have an amplitude that is high enough to produce meaningful output level, i.e. an output that is easy to measure accurately, but

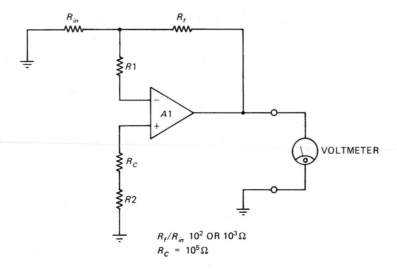

FIGURE 4-1 Input offset current test circuit.

low enough to not saturate the amplifier. When the input signal must be very low level, as might occur when the overall closed-loop gain is very high, a means must be provided to ensure that any DC input offset that is present does not affect the measurement. In those cases it is necessary to null the input offset using one of the methods discussed later in this chapter. In some cases these offsets can approach, or even exceed, the amplitude of the signal voltage.

4-3.2 Input offset current

Input offset current is measured in a test circuit such as shown in Fig. 4-1. Input offset current can be specified by the relationship between two different output offset voltages that are taken under different input conditions:

$$I_{io} = \frac{R_{in}(V_{o1} - V_{o2})}{R2(R_{in} + R_f)} \qquad (4\text{-}2)$$

The first output voltage, V_{o1}, is measured with $R1$ and $R2$ connected in the circuit. Voltage V_{o2} is then measured with resistors short-circuited, but all other conditions remaining the same. The resultant output voltage can, along with V_{o1}, be used in Eq. (4-2) to determine input offset current.

4-3.3 Input offset voltage

The input offset voltage is the voltage required to force the output voltage (V_o) to zero when the input voltage is also zero. The operational amplifier is connected in an inverting amplifier configuration such as Fig. 4-2. To make

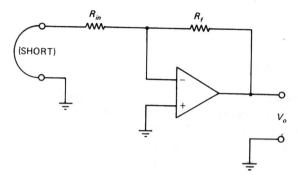

FIGURE 4-2   Input offset voltage test circuit.

the measurement, the input terminal is connected to ground. The input offset voltage is found by measuring the output voltage (when $V_{in} = 0$) and then using the standard voltage divider equation:

$$V_{io} = \frac{V_o R_{in}}{R_f + R_{in}} \tag{4-3}$$

Greater accuracy is achieved if the gain of the amplifier is 100, 1000 (or even higher), provided that such gains can be accommodated without saturating the amplifier.

4-3.4   Input bias current

This test requires a pair of closely matched resistors connected between the op-amp inputs (−IN and +IN) and ground (Fig. 4-3). Power is applied to the operational amplifier, and the voltage is measured at each input. The value of resistors $R1$ and $R2$ must be high enough to create a measurable voltage drop at the level of current anticipated. Although the actual value of these resistors is not too critical, the match between them ($R1 = R2$) is critical to the success of the measurement. The definition of 'readable' voltage drop depends upon the instrumentation available to do the job. For example, assume that an actual input bias current of five microamperes (5 μA) is flowing, and the resistors are each 10 kohms. In this case the measured voltage will be (by

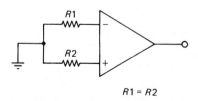

R1 = R2

FIGURE 4-3   Input bias current test circuit.

Ohm's law):

$$V = IR$$

$$V = (5 \times 10^{-6} \text{ A}) \times (10\,000\ \Omega)\ V = 5 \times 10^{-2}\ V = 50\ \text{mV}$$

If your voltage measuring equipment is not capable of measuring these levels, then higher resistor values would be required. If the two inputs had ideally equal input bias currents, only one measurement would be needed. Since real devices usually have unequal bias currents, however, it is sometimes necessary to measure both and use the higher input bias current this measured. Alternatively, the root sum squares (RSS) value can be used.

4-3.5 Power supply sensitivity (PSS)

Power supply sensitivity (ψ) is the worst-case change of input offset voltage for a 1.0 Vdc change of *one* DC power supply voltage (either $V-$ or $V+$), with the other supply potential being held constant. The same test configuration is used to measure this parameter as was used to measure input offset voltage (Fig. 4-2). First, the two power supply voltages are set to equal levels and the input offset voltage is measured. One of the power supply voltages is then changed by precisely 1.00 Vdc and the input offset voltage is again measured. The power supply sensitivity (PSS) is given by:

$$\psi = \frac{\Delta V_{io}}{\Delta V_o} \tag{4-4}$$

EXAMPLE 4-1

If the operational amplifier has an initial offset voltage of 5.0 mV, and a value of 6.2 mV is found to exist after the $V+$ power supply voltage is changed from $+12$ Vdc to $+13$ Vdc. Calculate the power supply sensitivity for a $+1$ Vdc change in $V+$.

Solution

$$\psi = \frac{\Delta V_{io}}{\Delta V_o} \quad \text{(Eq. (4-4) restated)}$$

$$\psi = \frac{(6.2 - 5.0)\ \text{mV}}{1.0\ \text{V}}$$

$$\psi = \frac{1.2\ \text{mV}}{1.0\ V} = 1.2\ \text{mV/V} \qquad \blacksquare$$

The actual power supply sensitivity is the worst case when this measurement is made under four conditions: (1) $V+$ increased 1 Vdc; (2) $V+$ decreased 1 Vdc; (3) $V-$ increased 1 Vdc; and (4) $V-$ decreased 1 Vdc. The worst case of these four measurements is taken to be the true power supply sensitivity.

4-3.6 Slew rate

Slew rate is a measure of the operational amplifier's ability to shift between the two possible opposite output voltage extremes while supplying full output power to the load. This parameter is usually specified in terms of volts per unit of time (e.g. 30 V/μs).

A saturating squarewave is usually used to measure the slew rate of an operational amplifier. The squarewave must have a risetime that substantially exceeds the expected slew rate of the operational amplifier. The value of rise-time is found from examination of the leading edge of the output waveform on an oscilloscope while the input is overdriven by the squarewave. The time measured is that which is required for the output to slew from 10% of the final value to 90% of the final value. It must be noted that slew rate can be affected by gain, so the value at unity gain will not match either the slew rate under open-loop or very high gain closed-loop conditions. Once the switching time is known the slew rate (S_r) is closely approximated by:

$$S_r = \frac{(V+) + |(V-)|}{T_s} \qquad (4\text{-}5)$$

where:

S_r is the slew rate in volts per microsecond (V/μs)

$V+$ is the positive supply voltage

$|(V-)|$ is the absolute value of the negative supply voltage

T_s is the switching time in microseconds (μs)

Since most manufacturers specify slew rate for the open-loop configuration in their data sheets, we can use this relationship to approximate the switching times of specific operational amplifier 'digital' circuits.

It is possible to improve the closed-loop slew rate at any given gain figure through the use of appropriate *lag compensation* techniques (Fig. 4-4). Keep

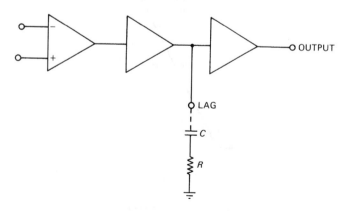

FIGURE 4-4   Lag frequency compensation.

the values of R_f and R_{in} low when trying to improve slew rates; values under 10 kohms will be best. The lag compensation capacitor will have a value of:

$$C = \left[\frac{R_f + R_{in}}{4\pi R_f R_{in}}\right]\left[\frac{F_{oi}}{10^m}\right] \tag{4-6}$$

where:

F_{oi} is the -3 dB half-power point

R_f is the feedback resistance

R_{in} is the input resistance

m is the quantity $[A_{vol(dB)} - A_{v(dB)}]/20$

The resistor value is found from:

$$R = \frac{1}{2\pi F_{oi} C} \tag{4-7}$$

It is the usual practice to measure slew rate in the noninverting unity gain voltage follower configuration because that circuit generally has the poorest slew rate in most op-amp devices. As in the previous test, the worst-case figure is used as a matter of standard practice.

The unity gain follower is driven by a squarewave of sufficient amplitude to drive the device well beyond the full saturation point. This criterion is necessary to eliminate the rounded curves which will exist at points just below full saturation. The output waveform can then be examined with a wideband oscilloscope which has a timebase fast enough to allow for a meaningful examination. The trace (Fig. 4-5) will be a straight line with a certain slope. It

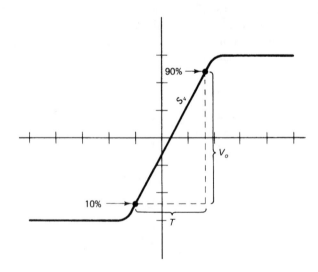

FIGURE 4-5 Slew rate is found by measuring the speed at which the output snaps from one extreme state to the other.

is standard practice to measure risetime as the time of transition from 10% of full amplitude to 90% of full amplitude. Adjust the timebase triggering of the oscilloscope so that the slope covers several horizontal scale divisions. The slew rate (S_r) is then found from the slope of the trace on the oscilloscope:

$$S_r = \frac{V_o}{t} \qquad (4\text{-}8)$$

4-3.7 Phase shift

The phase shift (ϕ) of an operational amplifier can be measured using a sinewave and an oscilloscope (CRO). In one version of this test an $X - Y$ oscilloscope (or a dual-channel model with an $X - Y$ capability) is used. The input signal is applied to the vertical channel of the CRO, while the operational amplifier output is applied to the horizontal channel. The gains for the two channels are set to produce equal beam deflections. The points marked $Y1$ through $Y4$ (Fig. 4-6A) are measured and the phase shift is calculated from:

$$\phi = \sin^{-1}\left[\frac{Y1 - Y2}{Y4 - Y3}\right] \qquad (4\text{-}9)$$

An alternative approach uses a dual-trace CRO in which the input signal is applied to one channel and the output signal is applied to the other channel. The noninverting unity gain operational amplifier configuration is used. The CRO channel gains are adjusted to be identical, and the traces are superimposed (Fig. 4-6B) on each other. The phase shift is found from:

$$\phi = \frac{360B}{A} \qquad (4\text{-}10)$$

EXAMPLE 4-2

In Fig. 4-6B the overall trace is adjusted to produce a horizontal deflection of 9.5 divisions on the CRO screen. When the traces are overlapped it is found that dimension B, the time difference between zero-crossings, is found to be 0.8 divisions. Calculate the phase shift.

Solution

$$\phi = \frac{360B}{A} \qquad \text{(Eq. (4-10) restated)}$$

$$\phi = \frac{(360)(0.8)}{9.5} = \frac{288}{9.5} = 30.32° \qquad \blacksquare$$

This test only works properly if both channels of the CRO are identical to each other, and thus have the same internal phase shift. In addition, the CRO must have a high internal chopping frequency in the dual beam mode. It must be recognized that instrument limitations often reduce the usefulness of this test.

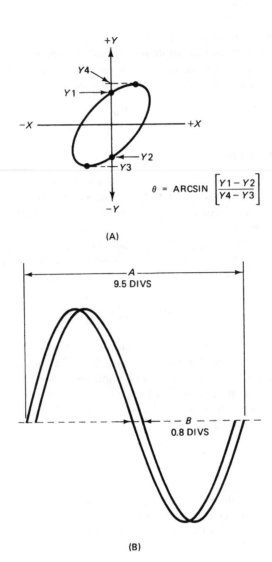

FIGURE 4-6   Phase measurement: (A) Lissajous pattern method; (B) timebase method.

4-3.8   Common mode rejection ratio (CMRR)

The common mode rejection ratio is defined as the ratio of the differential gain to the common mode gain:

$$CMRR = \frac{A_{vd}}{A_{vol}} \qquad (4\text{-}11)$$

where:

$CMRR$ is the common mode rejection ratio

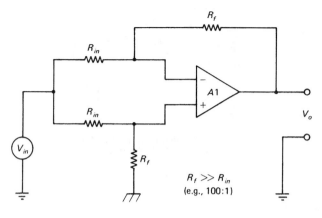

FIGURE 4-7   Common mode rejection test circuit.

A_{vd} is the differential voltage gain

A_{vcm} is the common mode voltage gain

The common mode voltage gain is not always a linear function of common mode input voltage (V_{cm}), so it is usually specified at the maximum allowable value for V_{cm}. The amplifier is connected into a circuit such as Fig. 4-7 and *CMRR* determined from:

$$CMRR = \frac{(R_f + R_{in})V_{in}}{R_{in}V_o}$$ (4-12)

provided that $R_f \gg R_{in}$, i.e. $\geq 100 : 1$.

4-4   DC ERRORS AND THEIR SOLUTIONS

In this section we will examine DC errors in operational amplifiers. These errors are in the form of current and voltage levels that conspire to force the output voltage to differ from the theoretical value in any given case. In many cases the error will be specified in terms of an actual output voltage that is non-zero at a time when it should be zero (e.g. when $V_{in} = 0$). We will also propose some circuit tactics that minimize or eliminate certain errors.

4-4.1   Output offset compensation

The DC error factors result in an output offset voltage V_{oo} which exists between the output terminal and ground at a time when V_o should be zero. This phenomena helps explain discrepancies between those voltages which actually are found to exist and those that the equations say should exist. One method for classifying output offset voltages is by the causes: *input offset voltage* and *input bias current*.

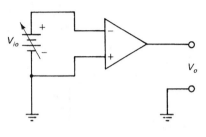

V_{OFF} = INPUT VOLTAGE REQUIRED
TO FORCE V_o TO ZERO

(A)

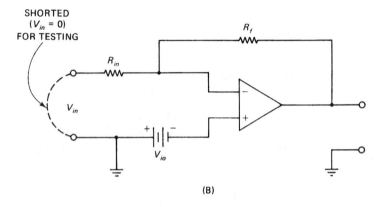

(B)

FIGURE 4-8   (A) Input offset voltage measurement circuit; (B) equivalent circuit model.

Input offset voltage V_{io} (Fig. 4-8A) is defined as the differential input voltage required to force the output to zero when no other signal is present ($V_{in} = 0$). A reasonably good model for the input offset voltage phenomenon (Fig. 4-8B) is a voltage source with one end connected to ground and the other end connected to the noninverting input. Although voltage source polarity is shown here, the actual polarity found in any given situation may be either positive or negative to ground, depending upon the device being tested. Values for input offset are typically from one to several millivolts. The popular type 741 operational amplifier is specified to have a 1 to 5 mV input offset voltage, with 2 mV being listed in the data sheet as typical.

The value of the output offset voltage (V_{oo}) caused by an input offset voltage is given by:

$$V_{oo} = R_f V_{io} \qquad (4-13)$$

If the circuit gain is low, and V_{in} remains at relatively high values, then the input offset voltage may be of little practical consequence. It is primarily where either high values of A_v, or low values of V_{in}, are encountered that the input offset voltage becomes a problem (these conditions are often encountered together).

The second major cause of output offset voltage is spurious input current. This can be further subdivided into two more classes: *normal input bias* and *input offset current*.

Figure 4-9A shows a typical operational amplifier differential input stage. Whenever bipolar (NPN or PNP) transistors are used in this stage some small input bias current will be required for normal operation. This is one

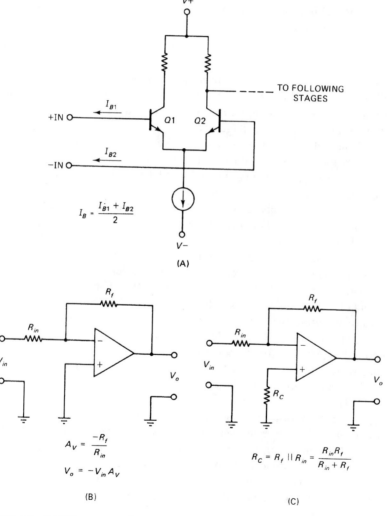

(A)

(B)

$$A_v = \frac{-R_f}{R_{in}}$$

$$V_o = -V_{in} A_v$$

(C)

$$R_c = R_f \parallel R_{in} = \frac{R_{in} R_f}{R_{in} + R_f}$$

FIGURE 4-9 (A) Differential input circuit for op-amp; (B) op-amp inverting follower circuit; (C) use of compensation resistor (R_c) to cancel output offset.

of those unavoidable conditions inherent in the nature of the transistor rather than any deficiency in the internal op-amp circuitry. The problem that input bias currents cause becomes acute when high values of input and feedback resistances are used (Fig. 4-9B). When these resistors are in the circuit the bias current causes a voltage drop across the resistances even when $V_{in} = 0$, causing an output voltage equal to:

$$V_{oo} = I_b \times R_f \qquad (4\text{-}14)$$

Figure 4-9C shows the use of a compensation resistor, R_c, to reduce the offset potential due to input bias currents. This same resistor also improves thermal drift. This resistor has a value equal to the parallel combination of the other two resistors:

$$R_c = \frac{R_f R_{in}}{R_f + R_{in}} \qquad (4\text{-}15)$$

Approximately the same bias current flows in both inputs, so the compensation resistor will produce the same voltage drop at the noninverting input as appears at the inverting input. Because the operational amplifier inputs are differential, the net output voltage is zero.

The method of Fig. 4-9C is used where the source of the DC offset potential in the output signal is due to the input bias currents.

If the input signal contains an undesired DC component, then the DC component will also create an amplifier output error. Depending upon the specific situation, we may need considerably greater range of control than is offered by Fig. 4-9C. For this purpose we turn to one of several external offset null circuit techniques.

Figure 4-10 shows two methods for nulling output offsets, regardless of whether the source is bias currents, other op-amp defects, or an input signal DC component. Figure 4-10A shows the use of the offset null terminals that are found on some operational amplifiers. A potentiometer is placed between the terminals, while the wiper is connected to the $V-$ power supply. This potentiometer is adjusted to produce the null required; the input terminals are shorted together at ground potential, and the potentiometer is adjusted to produce 0 Vdc output. In the LM-101/201/301 family of devices the wiper terminal of the potentiometer is connected to common rather than $V-$ (see Fig. 4-10B).

Figure 4-10C shows a nulling circuit that can be used on any operational amplifier, inverting or noninverting, except the unity gain noninverting follower. A counter-current ($I3$) is injected into the summing junction (point A) of a magnitude and polarity such as to cancel the output offset voltage. The voltage at point B is set to produce a null offset at the output. The output voltage component due to this voltage is:

$$V'_o = -V_B \times \frac{R_f}{R2} \qquad (4\text{-}16)$$

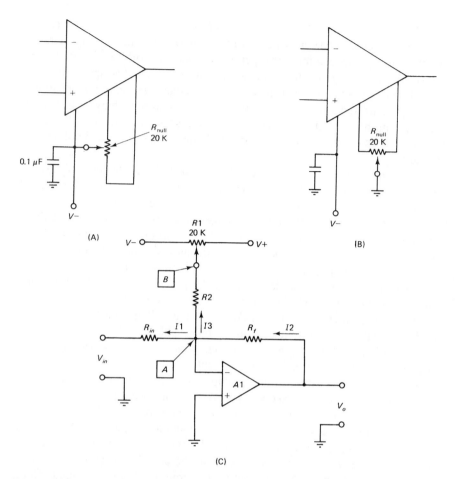

FIGURE 4-10 Alternative offset null methods: (A) use of offset null terminals on op-amp; (B) alternative method for using offset null terminals; (C) use of offset null potentiometer to inverting input.

If finer control over the output offset is required, then the potentiometer can be replaced with another resistor voltage divider consisting of a low-value potentiometer (e.g. 1 kohm) in series with two fixed resistors, one connected to each end of the potentiometer. Thus, the potentiometer resistance is only a fraction of the total voltage divider resistance. In some cases, narrower limits are set by using zener diodes or three-terminal IC voltage regulators at each end of the potentiometer to reduce the range of $V-$ and $V+$.

4-4.2 Thermal drift

Another category of DC error is *thermal drift*. This parameter is usually given in the data sheets relative to input conditions. The specification is usually related to the *change of input offset voltage per degree Celcius of temperature change*. Typical figures for common operational amplifiers are in the range 1 to 5 microvolts per degree Celcius (μV/°C), with typical values around 3 μV/°C. Keep in mind, however, that this specification is usually an expression of drift in steady state circuits and may not accurately represent drift under dynamic situations. These conditions may well be a lot higher than the drift given in the specification sheet for a device. There are actually two sources of drift in the input circuit; both current and voltage are at fault.

A potential source of increased thermal drift is in the use of resistor networks to null offset voltages (previous section and Fig. 4-10C). The current passing through the network adds to the drift because, as the resistors change value in response to temperature changes, the op-amp sees the resultant varying current as a valid input signal. The overall drift performance is thus made worse. Use of low temperature coefficient resistors helps this situation. In some cases, an op-amp manufacturer will recommend an offset null circuit that is different from those described in the previous section. Such circuits are preferred because they are generally designed to temperature compensate the process for a specific device. Although the null methods shown previously will work well for nulling, and are reasonable in stable temperature (or noncritical) situations, they may fall down in cases where temperature varies because they unbalance input offset currents.

The input offset current of an op-amp can result in an offset voltage. We discussed this problem in Section 4-1. A solution to the problem shown in that discussion is the use of a *compensation resistor* between the noninverting input and ground. That resistor has a value equal to the parallel combination of feedback and input resistances. Although couched in terms of the inverting follower circuit in the previous discussion, the actual goal is to ensure that both inputs (+IN and −IN) look out to the same resistance to ground. Thus, the differential nature of the op-amp inputs tends to cancel out the effects of the voltage drop created by bias currents flowing in external resistances.

The same method also works to improve thermal drift. The bias current is a function of temperature, hence thermal drift is due to bias current changes as a function of temperature change. These effects are smoothed out when both inputs see the same resistance. It is sometimes important to use identical, low temperature coefficient resistors in the op-amp input and feedback circuits. In other cases, thermistors or PN junction devices are used to temperature compensate circuits.

Another means for reducing overall thermal drift is to reduce thermoelectric sources in the circuit. A phenomena called the *Seebeck effect* exists when dissimilar metals are joined together. If two metals with different quantum work functions are joined together in a junction, then a voltage is generated between them that is proportional to the junction temperature. The Seebeck effect is used to form temperature sensors called *thermocouples*. Ordinary copper wire or printed wiring board tracks produces thermoelectric voltages of 40 to 60 μV/°C when brought into contact with the *Kovar*® leads used on integrated circuits. Lead/tin solder reduces the effect to 1 to 5 μV/°C, and cadmium-based solders to less than 0.5 μV/°C.

For very low level signals, especially where high gain is also present, several steps are taken to reduce thermal drift. First, operate the device at the lowest possible internal junction temperature. Two methods are useful in this respect. One is to use a heatsink (where possible) on the IC device. The other is to operate the temperature sensitive stage at the lowest practical DC power supply voltages; for example at ±6 Vdc instead of ±15 Vdc. Second, ensure that both amplifier inputs see the same resistance to ground (use a compensation resistor). Third, make all resistors in the circuit low temperature coefficient types. Fourth, eliminate thermoelectric sources. Fifth, stabilize the ambient temperature. Finally, if necessary use thermistor or PN junction devices to temperature compensate the circuit.

Although drift may not be too important in some circuits, it can become critical where low-level signals are being processed. If an amplifier has a drift of 10 μV/°C, and the circuit is expected to maintain its performance over a large temperature range, then a significant drift component may exist when low input signals are processed. For example, a temperature change of 20°C will result in a 200 μV error, or in other terms, an error that is 20 percent of a 1 mV input signal.

4-5 NOISE, SIGNALS AND AMPLIFIERS _____

Although gain, bandwidth and the shape of the passband are important amplifier characteristics, we must also be concerned about circuit *noise*.

At any temperature above absolute zero (0K or about −273°C) electrons in any material are in constant random motion. Because of the inherent randomness of that motion, however, there is no detectable current in any one direction. In other words, electron drift in any single direction is cancelled over short times by equal drift in the opposite direction. There is, however, a continuous series of random current pulses generated in the material, and those pulses are seen by the outside world as a noise signal. This signal is called by several names: *thermal agitation noise, thermal noise* or *Johnson noise*.

Johnson noise is a so-called 'white noise' because it has a very broadband Gaussian spectral density. The term white noise is a metaphor developed from white light, which is composed of all visible color frequencies. The expression for Johnson noise is:

$$V_n^2 = 4KTBR \text{ volts}^2/\text{Hz} \tag{4-17}$$

where:

V_n is the noise potential in volts (V)

K is Boltzmann's constant (1.38×10^{-23} J/K)

T is the temperature in Kelvin (K)

R is the resistance in ohms (Ω)

β is the bandwidth in hertz (Hz)

With the constants collected, and the expression normalized to 1 kohm, Eq. (4-17) reduces to:

$$V_n = 4\sqrt{\frac{R}{1\text{ k}\Omega}} \quad \text{nV}/\sqrt{\text{Hz}} \tag{4-18}$$

The evaluated solution of Eq. (4-18) is normally read *nanovolts of AC noise per squareroot hertz.*

EXAMPLE 4-3

Calculate the spectral noise for a 1 MΩ resistor.

Solution

$$V_n = 4\sqrt{\frac{R}{1\text{ k}\Omega}} \quad \text{nV}/\sqrt{\text{Hz}} \quad \text{(Eq. (4-18) restated)}$$

$$V_n = (4)\frac{\sqrt{1000\text{ k}\Omega}}{1\text{ k}\Omega} \quad \text{nV}/\sqrt{\text{Hz}}$$

$$V_n = (4)(31.6) \quad \text{nV}/\sqrt{\text{Hz}} = 126.4 \quad \text{nV}/\sqrt{\text{Hz}} \qquad \blacksquare$$

Several other forms of noise are present in linear ICs to one extent or another. For example, because current flow at the quantum level is not smooth and predictable, an intermittent noise burst phenomenon is sometimes seen. This noise is called 'popcorn noise' and consists of pulses of many milliseconds duration. Another form of noise is 'shot noise' (also called 'Schottky noise'). The name shot is derived from the fact that the noise sounds like a handful of B-B shot thrown against a metal surface. Shot noise is a consequence of DC current flowing in any conductor, and is found from:

$$I_n^2 = 2qIB \quad \text{A}^2/\text{Hz} \tag{4-19}$$

where:

I_n is the noise current in amperes (A)

q is the elementary electric charge (1.6×10^{-19} coulombs)

I is the current in amperes (A)

B is the bandwidth in hertz (Hz)

Finally, we see 'flicker noise', also called 'pink noise' or '1/F noise'. The latter name applies because flicker noise is predominantly a low frequency (<100 Hz) phenomenon. This type of noise is found in all conductors, and becomes important in IC devices because of manufacturing defects.

Amplifiers are evaluated on the basis of *signal-to-noise ratio* (S/N or SNR). The goal of the designer is to enhance the SNR as much as possible. Ultimately, the minimum signal detectable at the output of an amplifier is that which appears above the noise level. Therefore, the lower the system noise, the smaller the *minimum allowable signal.* Although usually thought of as a radio receiver parameter, SNR is applicable in other amplifiers where signal levels are low and gains are high. This situation occurs in scientific, medical and engineering instrumentation, as well as other applications.

Noise resulting from thermal agitation of electrons is measured in terms of *noise power* (P_n), and carries the units of power (watts or its sub-units). Noise power is found from:

$$P_n = KTB \qquad (4\text{-}20)$$

where:
P_n is the noise power in watts (W)
K is Boltzmann's constant (1.38×10^{-23} J/K)
B is the bandwidth in hertz (Hz)

Notice in Eq. (4-20) that there is no center frequency term, only a bandwidth. True thermal noise is *Gaussian* or (near Gaussian) in nature, so frequency content, phase and amplitudes are equally distributed across the entire spectrum. Thus, in bandwidth limited systems, such as a practical amplifier or network, the total noise power is related only to temperature and bandwidth. We can conclude that a 200 Hz bandwidth centered on 1 kHz produces the same thermal noise level as a 200 Hz bandwidth centered on 600 Hz or some other frequency.

Noise sources can be categorized as either *internal* or *external.* The internal noise sources are due to thermal currents in the semiconductor material resistances. It is the noise component contributed by the amplifier under consideration. If noise, or S/N ratio, is measured at both input and output of an amplifier, the output noise is greater. The internal noise of the device is the difference between output noise level and input noise level.

External noise is the noise produced by the signal source, so it is (not surprisingly) often called *source noise.* This noise signal is due to thermal agitation currents in the signal source, and even a simple zero-signal input termination resistance has some amount of thermal agitation noise.

Figure 4-11A shows a circuit model showing that several voltage and current noise sources exist in an op-amp. The relative strengths of these noise sources, hence their overall contribution, varies with op-amp type. In an FET-input op-amp, for example, the current noise sources are tiny, but voltage noise sources are very large. On bipolar op-amps the exact opposite situation obtains.

All of the noise sources in Fig. 4-11A are uncorrelated with respect to each other, so one cannot simply add noise voltages; only noise *powers* can

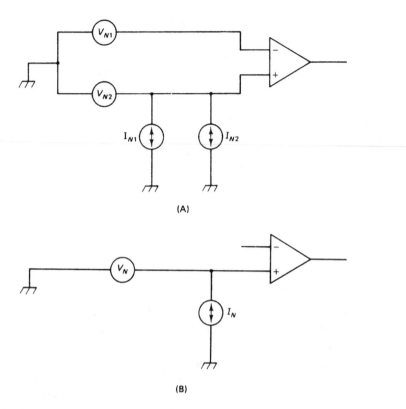

(A)

(B)

FIGURE 4-11 Current and voltage noise sources at amplifier input: (A) differential case; (B) single-ended case.

be added. To characterize noise voltages and currents they must be added in the root sum squares (RSS) manner.

Models such as Fig. 4-11A are too complex for most situations, so it is standard practice to lump all of the voltage noise sources into one source, and all of the current noise sources into another source. The composite sources have a value equal to the RSS voltage (or current) of the individual sources. Figure 4-11B is such a model in which only a single current source and a single voltage source are used. The *equivalent AC noise* in Fig. 4-11B is the overall noise, given a specified value of source resistance, R_s, and is found from the RSS value of V_n and I_n:

$$V_{\neq} = \sqrt{V_n^2 + (I_n R_s)^2} \qquad (4\text{-}21)$$

4-5.1 Noise factor, noise figure and noise temperature

The noise of a system or network can be defined in three different but related ways: *noise factor* (F_n), *noise figure* (NF) and *equivalent noise temperature*

(T_e); these properties are definable as a *ratio, decibel* or *temperature,* respectively.

Noise factor (F_n). The noise factor is the ratio of output noise power (P_{no}) to input noise power (P_{ni}):

$$F_n = \frac{P_{no}}{P_{ni}}\bigg|_{T=290K} \tag{4-22}$$

In order to make comparisons easier the noise factor is always measured at the standard temperature (T_o) 290K (approximately 'room temperature').

The input noise power P_{ni} can be defined as the product of the source noise at standard temperature (T_o) and the amplifier gain:

$$P_{ni} = GKBT_o \tag{4-23}$$

It is also possible to define noise factor F_n in terms of output and input S/N ratio:

$$F_n = \frac{SNR_{in}}{SNR_{out}} \tag{4-24}$$

which is also:

$$F_n = \frac{P_{no}}{KT_oBG} \tag{4-25}$$

where:

SNR_{in} is the input signal to noise ratio
SNR_{out} is the output signal to noise ratio
P_{no} is the output noise power in watts (W)
K is Boltzmann's constant (1.38×10^{-23} J/K)
T_o is 290 Kelvin (K)
B is the network bandwidth in hertz (Hz)
G is the amplifier gain

The noise factor can be evaluated in a model that considers the amplifier ideal, and therefore only amplifies through gain G the noise produced by the 'input' noise source:

$$F_n = \frac{KT_oBG + \Delta N}{KT_oBG} \tag{4-26}$$

or

$$F_n = \frac{\Delta N}{KT_oBG} \tag{4-27}$$

where ΔN is the noise added by the network or amplifier. All other terms are as defined above.

Noise figure (NF). The noise figure is a frequently used measure of an amplifier's 'goodness' or its departure from 'idealness'. Thus, it is a figure of merit. The noise figure is the noise factor converted to decibel notation:

$$NF = 10 \log F_n \qquad (4\text{-}28)$$

where:
NF is the noise figure in decibels (dB)
F_n is the noise factor
log refers to the system of base-10 logarithms

Noise temperature (T_e). The noise temperature is a means for specifying noise in terms of an equivalent temperature. Evaluating Eq. (4-23) shows that the noise power is directly proportional to temperature in Kelvin, and also that noise power collapses to zero at the temperature of absolute zero (0K).

Note that the equivalent noise temperature T_e is *not* the physical temperature of the amplifier, but rather a theoretical construct that is an *equivalent* temperature that produces that amount of noise power. The noise temperature is related to the noise factor by:

$$T_e = (F_n - 1)T_o \qquad (4\text{-}29)$$

and to noise figure by:

$$T_e = \left[\text{antilog} \left(\frac{NF}{10} \right) \right] KT_o \qquad (4\text{-}30)$$

Now that we have noise temperature T_e, we can also define noise factor and noise figure in terms of noise temperature:

$$F_n = \left(\frac{T_e}{T_o} \right) - 1 \qquad (4\text{-}31)$$

and

$$NF = 10 \ \log \left[\left(\frac{T_e}{T_o} \right) + 1 \right] \qquad (4\text{-}32)$$

The total noise in any amplifier or network is the sum of internally generated and externally generated noise. In terms of noise temperature:

$$P_{n(total)} = GKB(T_o + T_e) \qquad (4\text{-}33)$$

where $P_{n(total)}$ is the total noise power and all other terms are as previously defined.

4-5.2 Noise in cascade amplifiers

A noise signal is seen by a following amplifier as a valid input signal. Thus, in a cascade amplifier the final stage sees an input signal that consists of the original signal and noise amplified by each successive stage. Each stage in the cascade chain both amplifies signals and noise from previous stages, and also contributes some noise of its own. The overall noise factor for a cascade amplifier can be calculated from *Friis' noise equation*:

$$F_n = F_1 + \frac{F_2 - 1}{G1} + \frac{F_3 - 1}{G1\,G2} + \cdots + \frac{F_n - 1}{G1\,G2 \cdots G_{n-1}} \qquad (4\text{-}34)$$

where:

F_n is the overall noise factor of N stages in cascade
F_1 is the noise factor of stage-1
F_2 is the noise factor of stage-2
F_n is the noise factor of the nth stage
$G1$ is the gain of stage-1
$G2$ is the gain of stage-2
G_{n-1} is the gain of stage $(n - 1)$

As you can see from Eq. (4-34), the noise factor of the entire cascade chain is dominated by the noise contribution of the first stage or two. Typically, high gain amplifiers use a low noise device for only the first stage or two in the cascade chain.

4-6 AC FREQUENCY STABILITY

Operational amplifiers are subject to spurious oscillations, especially those that are not internally frequency compensated. Figure 4-12 shows the plot of

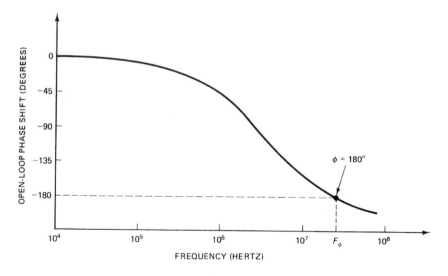

FIGURE 4-12 Open-loop phase shift versus frequency.

the open-loop phase shift versus frequency for a typical operational amplifier. From DC to a certain frequency there is essentially zero phase-shift error, but above that breakpoint the phase error increases rapidly. This change is due both to the internal resistances and capacitances of the amplifier acting as an *RC* phase-shift network, and the phase shift of the feedback network. This shift is called the propagation phase shift. At some frequency, *F*, the propagation phase-shift error reaches 180°, which when added to the 180° inversion that is normal in an inverting follower amplifier, adds up to the 360° phase shift that satisfies *Barkhausen's criteria* for oscillation. At this frequency the amplifier will become an oscillator.

In some cases, we may need to use a variant of the methods shown in Fig. 4-13. In Fig. 4-13A we see lead compensation. If the operational amplifier is equipped with compensation terminals (usually either pins 1 and 8, or 1 and 5, on 'standard' packages), then connect a small value capacitor (20 to 100 pF) as shown. An alternative scheme is to connect the capacitor from a compensation terminal to the output terminal.

The recommended capacitance in manufacturer's specification sheets is for the unity gain noninverting follower configuration. For a gain follower the capacitance is reduced by the feedback factor, β:

$$C = C_m \beta \qquad (4\text{-}35)$$

where:

    C is the required capacitance
    C_m is the recommended unity gain capacitance
    β is the feedback factor $R_{in}/(R_{in} + R_f)$

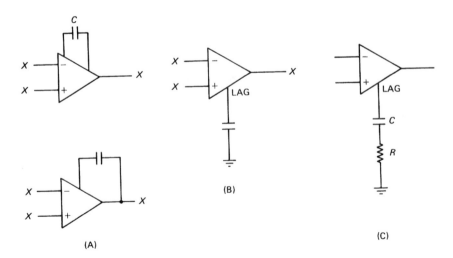

FIGURE 4-13 Frequency compensation methods.

Lag compensation is shown in Figs 4-13B and 4-13C. In this case, either a single capacitor (Fig. 4-13B) or a resistor–capacitor network (Fig. 4-13C) is connected from the compensation terminal to ground. A related method places the resistor–capacitor series network between the inverting and noninverting input terminals.

The object of these methods (see Fig. 4-14) is to reduce the high frequency loop gain of the circuit to a point where the total loop gain is less than unity at the frequency where the 180° phase shift occurs. The amount of compensation required to accomplish this goal determines the maximum amount of feedback that can be used without violating the stability requirement.

There are several factors that conspire toward allowing an operational amplifier to oscillate at times when this is highly undesirable. Quite often these oscillations occur at frequencies far in excess of the passband of the associated circuit. Two of these factors, both of which can be overcome, are positive feedback via the DC power supply and spurious internal phase shift.

Figure 4-15 shows the input and output waveforms from a unity gain inverting follower amplifier ($R_{in} = R_f$) that used no decoupling of $V-$ and $V+$ terminals. The output waveform shows a high frequency oscillation superimposed on the output sinewave. The cause of this problem is a high AC impedance to ground through the DC power supply terminals. The use of power supply decoupling helps somewhat for this problem, and it is considered poor engineering practice to use an uncompensated operational amplifier without those decoupling capacitors.

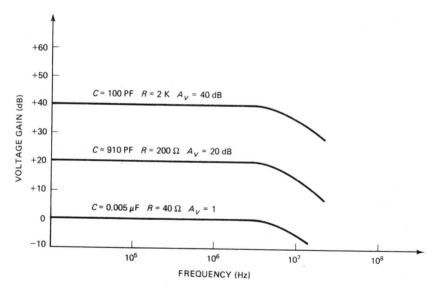

FIGURE 4-14 Voltage gain frequency response curves for three different combinations of resistance (R) and capacitance (C).

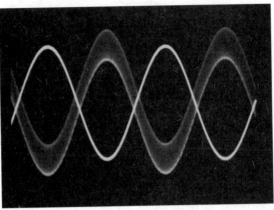

FIGURE 4-15 Input and output waveforms of amplifier with no frequency decoupling on V– and V+ power supply lines.

4-6.1 Diagnosing and fixing instability

The process of finding and fixing instability sources in linear IC amplifier circuits is sometimes believed (erroneously) to rely a little on magic, but in reality the process is relatively straightforward. There may be, however, a certain amount of cut-and-try involved.

But first, be sure that the problem that you are trying to solve is indeed an instability in the circuit, and not an external problem. Three kinds of external problem can mimmick op-amp instability. One is a defective source signal. For example, a low-frequency (sub-hertzian) 'motorboating' oscillation may be due to a high AC impedance to ground in the amplifier DC power supply, but it may also be due to an oscillation in the signal source. Another problem of this sort is 50/60 Hz AC hum caused by a broken shield or ground wire on the input line. In fact, don't look for any other cause if the 'oscillation' is exactly 50/60 Hz (the AC power line frequency) until this possibility is checked.

The second external problem is noisy or defective power supply lines. Examine the V– and V+ lines on an oscilloscope to be sure that the power is clean. A 120 Hz 'oscillation' could indicate an excessive ripple condition in a DC power supply. Additional ripple filtering or a voltage regulator in each DC line (V– and V+) are the usually successful solutions.

Finally, we sometimes see electromagnetic interference in IC amplifiers. Strong local fields, such as RF interference from a local broadcasting station, may get into the circuit and appear like high frequency oscillations. Such interference may be direct radiation, or may be carried into the circuit on an input line, output line, or a DC power supply line. Figure 4-16 shows an op-amp circuit containing several anti-interference techniques. The small-value capacitors should be used with caution because they may attenuate signals

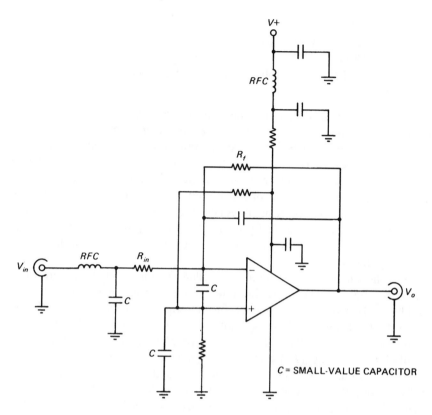

FIGURE 4-16   Several methods employed to deal with internal instability and external electromagnetic interference (EMI).

at higher frequencies. Also, they may add to the phase shift of signals in the feedback network and thereby cause an oscillation condition even while eliminating the interference.

Once the obvious non-instabilities have been eliminated, it is necessary to home in on possible causes of actual instabilities. First, examine the circuit for any *unintended common impedances* between input and output circuits, or between stages in a cascade chain of amplifiers. Two areas of concern are often found: the DC power supply lines and ground. The power supply circuits are usually handled by the bypass capacitors discussed earlier.

Common impedances in the ground system, usually called 'ground loops', are suppressed by using large conductive ground planes (usually on the component side of the printed circuit board) rather than thin point-to-point conductors. Diagnosis and resolving ground loops is subject to design and a few rules, beyond which a cut-and-try approach is advised. The cut-and-try method involves moving ground connections or actual components around the board to find 'cold' spots on the ground plane. Some rules to follow are:

1. Use large ground plane surfaces where possible.
2. Use single-point 'star' grounding rather than either random grounding or daisy-chain grounding.
3. Keep the DC power supply and AC signal lines separate except at the single star point.

Another source of difficulty, especially (but not exclusively) in high gain circuits, is erroneous circuit layout. Keep input and output circuits separated to the extent possible. A source of unintended feedback is sometimes found when several stages are in cascade. In response to printed wiring board size constraints some designers double a circuit back on itself on the board. That arrangement places the output circuit components next to either the input circuit components, or intermediate stages. Such a layout induces radiation feedback, and is especially likely if the total propagation feedback is $360°$ (e.g. as in noninverting amplifiers, or circuits with an even number of inverting amplifiers).

Once layout problems, common impedances, and other causes are ruled out, it is time to evaluate the linear IC amplifier circuit itself for instability problems. Two distinct areas must be investigated: feedback network and the rest of the circuit. Although not an absolute indicator, the frequency of oscillation often tells you where the problem is located. Measure the oscillation frequency (F_o) and compare it to the frequency at which the amplifier gain drops to unity (F_t) (which is found in the specifications sheet or data book). If F_o is close to F_t, then it is probable that the problem is in the feedback network. A diagnostic aid is to temporarily increase the amplifier closed-loop gain by a factor of $\times 2$ to $\times 10$, and then observe the effect on stability. If the oscillation ceases, or F_o drops appreciably, then the problem is probably in the feedback network. If neither of these events occurs, then the problem is most likely in another part of the circuit (the feedback loop is exonerated).

A few linear IC devices (including op-amps) operate as amplifiers into the VHF region, yet even low frequency op-amps and other low frequency devices sometimes *oscillate* in the 50 to 200 MHz region, way beyond the bandwidth of the device. The cause of these VHF parasitic oscillations is the output power amplifier in the IC device. The effect is especially likely when resonances are present.

One source of a stray resonance that leads to parasitics is the use of the wrong capacitor type on the DC power supply terminals. In Fig. 4-17A, for example, we see disk ceramic capacitors used for $V-$ and $V+$ bypassing. Such capacitors have significant stray capacitance and inductance (see inset) that tend to resonate in the VHF region. Fixes for test problems are shown in Fig. 4-17B. The $V+$ lead bypass capacitor uses a 2 to 12 ohm 'snubber' resistor in series to lower the Q of the stray resonant LC circuit elements. If the Q is lowered sufficiently, then the LC circuit elements are unable to cause

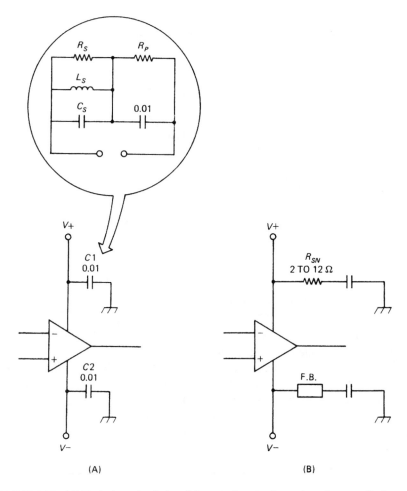

FIGURE 4-17 (A) Equivalent circuit for disk ceramic capacitors show how oscillation can occur at some VHF or UHF frequencies; (B) methods for dealing with capacitor resonances include snubber resistors and ferrite beads.

an oscillation. In the $V-$ lead of the same op-amp we see an alternative fix. A ferrite bead is slipped over the lead to the bypass capacitor. These beads act like RF chokes at VHF frequencies, but are practically transparent to low frequencies.

An often hidden source of feedback problems is capacitance in the load of the amplifier. Such capacitance adds to the propagation phase shift of the feedback network, possibly causing oscillation. If a load is known to be capacitive, then identification of the problem is easy. But it often happens that other sources of capacitance are found in a circuit. For example, shielded or coaxial cables have a high value of capacitance per unit of length. Similarly,

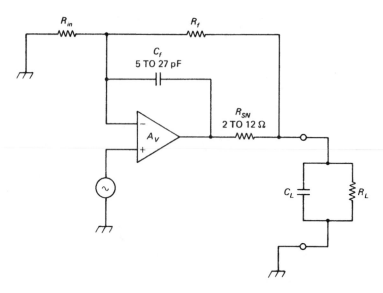

FIGURE 4-18   Circuit to isolate capacitive load from the feedback network.

some chassis or in-line connectors offer significant capacitances, and circuit stray capacitance may also be significant.

A circuit fix that isolates a capacitance load from an IC amplifier output is shown in Fig. 4-18. A small feedback capacitor reduces the closed-loop gain at frequencies where oscillation is likely to occur, while essentially not affecting lower frequencies at all. Isolation is obtained by the series snubber resistor, R_{sn}.

4-7   SUMMARY

1. Real operational amplifiers diverge considerably from the ideal devices discussed in textbooks. In many cases, the specifications seem nearly ideal, but actually depart quite a bit from the ideal when test conditions are considered.

2. Slew rate is a measure of an op-amp's ability to shift between two possible opposite output voltage extremes while supplying full output power to the load. This parameter is usually expressed in volts per microsecond (V/μs).

3. DC errors include input offset voltages, input offset currents and output offset voltages. These errors can be corrected by using a compensation resistor, a potentiometer attached to the offset null terminals of the device, or an external universal null circuit.

4. Thermal drift is a change in the input offset voltage per degree Celcius of temperature change. Use of a compensation resistor reduces thermal drift.

5. Noise generated inside the op-amp and in its external circuitry obscures weak signals. Low-noise operation of a cascade chain is possible if the input stage is of low-noise design (see Friis' equation).

6. Oscillation in an operational amplifier occurs when the propagation phase shift and any inherent inversion phase shift add up to $360°$ at any frequency where loop gain is unity or more. Oscillation can be overcome by compensating the circuit with a capacitor or RC network that reduces the gain at higher frequencies.

7. Oscillation can often be diagnosed by examining the frequency. If the frequency of oscillation is close to the natural gain–bandwidth product of the amplifier, then the problem is most likely in the feedback loop. Otherwise, look elsewhere for the problem.

4-8 RECAPITULATION

Now return to the objectives and Pre-quiz questions at the beginning of the chapter and see how well you can answer them. If you cannot answer certain questions, place a check mark to each and review the appropriate parts of the text. Next, try to answer the questions and work the problems below, using the same procedure.

4-9 QUESTIONS AND PROBLEMS

1. If the operational amplifier has an initial offset voltage of 4.0 mV, and a value of 5.6 mV is found to exist after the $V+$ power supply voltage is changed from $+12$ Vdc to $+13$ Vdc. Calculate the power supply sensitivity for a $+1$ Vdc change in $V+$.

2. If the operational amplifier has an initial offset voltage of 5.0 mV, and a value of 12 mV is found to exist after the $V+$ power supply voltage is changed from $+14$ Vdc to $+15$ Vdc. Calculate the power supply sensitivity for a $+1$ Vdc change in $V+$.

3. The trace on an oscilloscope is adjusted to produce a horizontal deflection of 8.5 divisions on the CRO screen. When the traces are overlapped it is found that dimension B, the time difference between zero-crossings, is found to be 0.96 divisions. Calculate the phase shift.

4. An oscilloscope is adjusted to produce a horizontal deflection of 8.8 divisions on the CRO screen. When the traces are overlapped it is found that dimension B, the time difference between zero-crossings, is found to be 0.86 divisions. Calculate the phase shift.

5. Calculate the spectral noise for a 150 kohm resistor at room temperature.

6. Calculate the spectral noise for a 10 megohm resistor: (a) at room temperature; (b) at an ambient temperature of $40°C$.

7. State Friis' equation for noise in a cascade amplifier and describe how it applies to creating a low-noise amplifier system.

8. An operational amplifier specification sheet lists a compensation capacitor of 47 pF in unity gain circuits.

9. Calculate the capacitance required in a noninverting follower with a gain of 11.

10. An operational amplifier specification sheet lists a compensation capacitor of 82 pF in unity gain circuits.

11. Calculate the capacitance required in a noninverting follower with a gain of 15.

12. An inverting follower amplifier uses a feedback resistor of 1.5 megohms and an input resistor of 100 kohms.

13. Calculate the value of a compensation resistor that will cancel input bias currents.

14. An inverting follower amplifier uses a feedback resistor of 150 kohms and an input resistor of 100 kohms.

15. Calculate the value of a compensation resistor that will cancel input bias currents.

16. An operational amplifier in a comparator configuration switches from $-V_{max}$ to $+V_{max}$ in 135 nanoseconds. Assume DC power supply voltages of ± 12 volts DC. What is the slew rate?

17. A DC differential amplifier has a voltage gain of 500, and a measured CMRR of 90 dB. Calculate the common mode voltage gain of this amplifier.

18. A 50 μA input offset current flows in an operational amplifier inverting input circuit in which the feedback resistor is 150 kohms, and the input resistor is 10 kohms. Calculate the output offset voltage due to this current.

19. Sketch the circuit for a universal output offset voltage null control. If the feedback resistor is 100 kohms, and the input resistor is 10 kohms, select values for the nulling circuit that will null the output voltage over a range of ± 10 volts.

20. An amplifier with a gain of 500 and a bandwidth of 1000 Hz is operated at room temperature. Calculate the output noise power of this circuit if the noise factor is 20.

21. Calculate the noise factor of an amplifier if the input S/N ratio is 4.5 dB and the output S/N ratio is 8.9 dB.

22. Calculate the noise added to the signal by an amplifier that has a noise factor of 10 when the bandwidth is 10 kHz and the gain is 200.

23. Calculate the equivalent noise temperature of an amplifier that has a noise factor of 4.5.

24. Calculate the noise figure of an amplifier that has an equivalent noise temperature of 270K.

25. A cascade chain of three amplifiers ($A1$, $A2$ and $A3$) has the following stage gain: $G1 = 40$, $G2 = 5$ and $G3 = 2$. Calculate the total noise factor if $NF1 = 3$, $NF2 = 7$ and $NF3 = 10$.

26. Discuss the most likely cause of oscillation in an operational amplifier circuit if the frequency of oscillation is near the gain–bandwidth product of the device.

CHAPTER 5

DC differential operational amplifier circuits

OBJECTIVES

1. Learn the properties of the DC differential amplifier.
2. Be able to design DC differential amplifier circuits.
3. Understand the range of applications for the DC differential amplifier.
4. Understand the limitations of the DC differential amplifier.

5-1 PRE-QUIZ

These questions test your prior knowledge of the material in this chapter. Try answering them before you read the chapter. Look for the answers (especially those you answered incorrectly) as you read the text. After you have finished studying the chapter try answering these questions again, and those at the end of the chapter (see Section 5-11).

1. Describe the difference between common mode gain and differential gain.
2. Solve the voltage gain transfer equation for a DC differential amplifier in which both input resistors are 10 kohms, and both feedback resistors are 200 kohms.
3. Define common mode rejection ratio (CMRR).
4. Draw the circuit for a single op-amp DC differential amplifier that incorporates a CMRR ADJUST control.

5-2 INTRODUCTION

Most operational amplifiers have differential inputs; that is, there are two inputs (−IN and +IN) that each provide the same amount of gain but have

135

opposite polarities. The inverting input (−IN) of the operational amplifier provides an output that is 180° out of phase with the input signal. In other words, a positive-going input signal will produce a negative-going output signal, and vice versa. The noninverting input (+IN) produces an output signal that is in-phase with the input signal. For this type of input, a positive-going input signal will produce a positive-going output signal. We will use these properties to understand a class of amplifiers in which both inputs are used. But before going further with our discussion of differential amplifiers, let's revisit the basic differential input stage of an operational amplifier.

5-3 DIFFERENTIAL INPUT STAGE

Figure 5-1 shows in simplified form a typical input stage for a differential input operational amplifier. Transistors $Q1$ and $Q2$ are a matched pair that

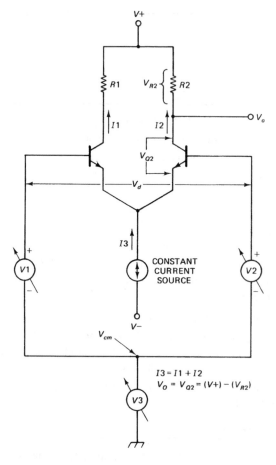

FIGURE 5-1 Differential amplifier input stage.

share common $V-$ and $V+$ DC power supplies through separate collector load resistors. The collector of transistor $Q2$ serves as the output terminal for the differential amplifier. Thus, collector voltage V_0 is the output voltage of the stage. The exact value of the output voltage is the difference between supply voltage $(V+)$ and the voltage drop across resistor $R2$ (i.e. V_{R2}).

The emitter terminals of the two transistors are connected together, and are fed from a constant current source (CCS). For purposes of analysis, we can assume that the collector and emitter currents ($I1$ and $I2$) are equal to each other (even though these currents are not actually equal — they are very close to each other — this convention is nonetheless useful for our present purpose). Because of Kirchhoff's current law (KCL), we know that:

$$I3 = I1 + I2 \qquad (5\text{-}1)$$

Because current $I3$ is a constant current, a change in either $I1$ or $I2$ will change the other. For example, an increase in current $I1$ must force a decrease in current $I2$ in order to satisfy Eq. (5-1). Now let's consider how the stage works. First, the case where $V1 = V2$. In this case, both $Q1$ and $Q2$ are biased equally, so the collector currents ($I1$ and $I2$) are equal to each other. Under the normal situation, this case will result in V_{R2} being equal to V_0, and also equal to $(V+)/2$.

Next, let's consider the case where $V1$ is greater than $V2$. In this case, transistor $Q1$ is biased on harder than $Q2$, so $I1$ will increase. Because $I1 + I2$ is always equal to $I3$, an increase in $I1$ will force a decrease in $I2$. A decrease of $I2$ will also reduce V_{R2} and increase V_0. Thus, an increase in $V1$ increases V_0, so we may conclude that the base of transistor $Q1$ is the noninverting input of the differential amplifier.

Now let's consider the case where $V2$ is greater than $V1$. In this case, transistor $Q2$ is biased on harder than $Q1$ so $I2$ will increase and $I1$ will decrease. An increase in $I2$ forces a larger voltage drop (V_{R2}) across resistor $R2$, so V_0 will go down. Thus, an increase in $V2$ forces a decrease in V_0, so the base of transistor $Q2$ is the inverting input of the differential amplifier.

Voltage $V3$ is a common mode signal, so will affect transistors $Q1$ and $Q2$ equally. For this kind of signal the output voltage (V_0) will not change.

The base bias currents required to keep transistors $Q1$ and $Q2$ operating become the input bias currents that make the practical op-amp non-ideal. In order to make the input impedance high we need to make these currents very low. Some manufacturers offer operational amplifiers with MOSFET transistors (called BiMOS op-amps) or JFET transistors (BiFET op-amps). These types of transistors inherently have lower input bias currents than bipolar NPN or PNP transistors, and in fact the bias currents approach mere leakage currents.

5-4 INPUT SIGNAL CONVENTIONS

Figure 5-2 shows a generic differential amplifier with the standard signals applied to the inputs. Signals $V1$ and $V2$ are single-ended potentials applied to the $-$IN and $+$IN inputs, respectively. The *differential input signal*, V_d, is the difference between the two single-ended signals: $V_d = (V2 - V1)$. Signal V_{cm} is a *common mode signal*, that is, it is applied equally to both $-$IN and $+$IN inputs. These signals are described further below.

Common mode signals. First let's consider the common mode signal, V_{cm}. A common mode signal is one which is applied to both inputs at the same time. Such a signal might be either a voltage such as V_{cm}, or, a case where voltages $V1$ and $V2$ are equal to each other and of the same polarity (i.e. $V1 = V2$). The implication of the common mode signal is that, being applied equally to inverting and noninverting inputs, the output voltage is zero. Because the two inputs have equal but opposite polarity gains for common mode signals, the net output signal in response to a common mode signal is zero.

The operational amplifier with differential inputs cancels common mode signals. An example of the usefulness of this property is in the performance of the differential amplifier with respect to 50/60 Hz hum pick-up from local AC power lines. Almost all input signal cables for practical amplifiers will pick up 50/60 Hz radiated energy and convert it to a voltage that is seen by the amplifier as a genuine input signal. In a differential amplifier, however, the 50/60 Hz field will affect both inverting and noninverting input lines equally, so the 50/60 Hz artifact signal will disappear in the output.

The practical operational amplifier will not exhibit perfect rejection of common mode signals. A specification called the *common mode rejection ratio* (CMRR) tells us something of the ability of any given op-amp to reject such signals. The CMRR is usually specified in decibels (dB), and is

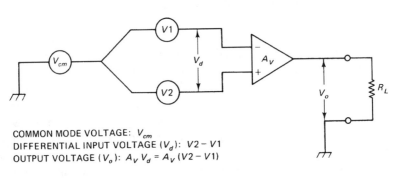

COMMON MODE VOLTAGE: V_{cm}
DIFFERENTIAL INPUT VOLTAGE (V_d): $V2 - V1$
OUTPUT VOLTAGE (V_o): $A_V V_d = A_V (V2 - V1)$

FIGURE 5-2 Signal voltages applied to op-amp differential inputs.

defined as:

$$CMRR = \frac{A_{vd}}{A_{cm}} \tag{5-2}$$

or, in decibel form:

$$CMRR_{dB} = 20 \log \frac{A_{vd}}{A_{cm}} \tag{5-3}$$

where:
$CMRR$ is the common mode rejection ratio
A_{vd} is the voltage gain to differential signals
A_{cm} is the voltage gain to common mode signals

In general, the higher the CMRR the better the operational amplifier. Typical low-cost devices have CMRR ratings of 60 dB or more, while premium devices offer CMRR of 120 dB or more.

EXAMPLE 5-1

An operational amplifier exhibits an open-loop differential gain of 100, and a common mode gain of 0.001. Calculate the CMRR in decibels.

Solution

$$CMRR_{dB} = 20 \log \left[\frac{A_{vd}}{A_{cm}} \right]$$

$$CMRR_{dB} = 20 \log \left[\frac{100}{0.001} \right] = (20)(5) = 100 \text{ dB} \qquad \blacksquare$$

Differential signals. Signals $V1$ and $V2$ in Fig. 5-2 are single-ended signals. The total differential signal seen by the operational amplifier is the difference between the single-ended signals:

$$V_d = V2 - V1 \tag{5-4}$$

The output signal from the differential operational amplifier is the product of the differential voltage gain and the difference between the two input signals (hence the term 'differential' amplifier). Thus, the transfer equation for the operational amplifier is:

$$V_o = A_v(V2 - V1) \tag{5-5}$$

EXAMPLE 5-2

A DC differential amplifier has a gain of 200. Calculate the output voltage if $V1 = 2.06$ Vdc and $V2 = 2.03$ Vdc.

Solution

$$V_o = A_v(V2 - V1)$$

$$V_o = (200)(2.03 - 2.06) \text{ Vdc}$$

$$V_o = (200)(-0.03) \text{ Vdc} = -6 \text{ Vdc}$$ ∎

5-5 DIFFERENTIAL AMPLIFIER TRANSFER EQUATION ——————

The basic circuit for the DC differential amplifier is shown in Fig. 5-3A. This circuit uses only one operational amplifier, so is the simplest possible configuration. In Chapter 6 you will see additional circuits based on two or three operational amplifier devices. In its most common form the circuit of Fig. 5-3A is balanced such that $R1 = R2$ and $R3 = R4$.

Consider the redrawn differential amplifier circuit shown in Fig. 5-3B. Assume that source resistances R_{S1} and R_{S2} are zero. Further assume that $R1 = R2 = R$, and that $R3 = R4 = kR$, where k is a multiplier of R.

1. Set $V2 = 0$. In this case V_o is found from:

$$V_{01} = \frac{-kR}{R} V1 \qquad (5\text{-}6)$$

$$V_{01} = -kV1 \qquad (5\text{-}7)$$

2. Now assume that $V1 = 0$ instead:

$$V_a = \frac{V2 \, kR}{kR + R} \qquad (5\text{-}8)$$

$$V_a = \frac{V2 \, k}{k + 1} \qquad (5\text{-}9)$$

$$V_{02} = \left[\frac{kR}{R} + 1 \right] V_a \qquad (5\text{-}10)$$

$$V_{02} = \left[\frac{kR}{R} + 1 \right] \left[\frac{V2 \, k}{k + 1} \right] \qquad (5\text{-}11)$$

$$V_{02} = (k + 1) \left[\frac{V2 \, k}{k + 1} \right] \qquad (5\text{-}12)$$

$$V_{02} = V2 \, k \qquad (5\text{-}13)$$

3. Now, by superimposing the two expressions for V_o:

$$V_o = V_{02} + V_{01} \qquad (5\text{-}14)$$

$$V_o = (V2 \, k) - (V1 \, k) \qquad (5\text{-}15)$$

$$V_o = K(V2 - V1) \qquad (5\text{-}16)$$

According to Eq. (5-16), therefore, the output voltage is the product of the difference between single-ended input potentials $V1$ and $V2$, and a factor k. The differential input voltage (V_d) is:

$$V_d = V2 - V1 \qquad (5\text{-}17)$$

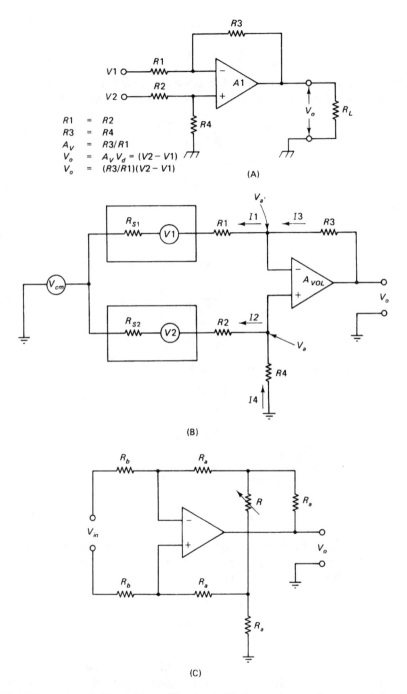

FIGURE 5-3   (A) DC differential amplifier circuit; (B) signal voltage and current relationships in the circuit; (C) one attempt to solve the gain control problem.

while factor k is the differential voltage gain, AV_{vd}. Thus, the output voltage is:

$$V_o = V_d A_{vd} \qquad (5\text{-}18)$$

We may also use a less parametric analysis by appealing to Fig. 5-3B. We know that the following relationships are true (assuming $R_{S1} = R_{S2} = 0$):

$$V_a = \frac{(V_{cm} + V2)R4}{R2 + R4} \qquad (5\text{-}19)$$

$$I1 = -I3 \qquad (5\text{-}20)$$

$$I1 = \frac{V_{cm} + V1 - V_{a'}}{R1} \qquad (5\text{-}21)$$

$$I3 = \frac{V_o - V_{a'}}{R3} \qquad (5\text{-}22)$$

From the properties of the ideal op-amp we know that voltage $V_{a'} = V_a$, so;

$$I1 = \frac{V_{cm} + V1 - V_a}{R1} \qquad (5\text{-}23)$$

$$I3 = \frac{V_o - V_a}{R3} \qquad (5\text{-}24)$$

Combining equations and solving for V_o:

$$V_0 = V_{cm}\left[\frac{R3R4 + R1R4 - R2R3 - R3R4}{R1(R2 + R4)}\right] - R3\left[\frac{V1}{R1}\right]$$
$$+ V2\left[\left(\frac{R4}{R2}\left(\frac{1 + R3/R1}{1 + R4/R2}\right)\right)\right] \qquad (5\text{-}25)$$

Assuming that $R1 = R2$ and $R3 = R4$, Eq. (5-25) resolves to:

$$V_o = \frac{R3}{R1}(V2 - V1) \qquad (5\text{-}26)$$

Equation (5-26) is similar to Eqs (5-5) and (5-16). In this case, A_{vd} is $R3/R1$ and V_d is $(V2 - V1)$. The standard transfer equation for the single op-amp DC differential amplifier is:

$$V_o = V_d A_{vd} \qquad (5\text{-}27)$$

$$V_o = V_d \frac{R3}{R1} \qquad (5\text{-}28)$$

It is difficult to build a DC differential amplifier with variable gain control. It is, for example, very difficult to get two ganged potentiometers (used to replace $R3$ and $R4$ in Fig. 5-3A) to track well enough to vary gain while maintaining required balance. Figure 5-3C shows one attempt to solve the problem. In this case, a potentiometer is connected between the midpoint of the two feedback resistances. This circuit works, but the gain control is a nonlinear function of the potentiometer setting. It is generally assumed to be a better practice to use a post-amplifier stage following the differential amplifier, and perform gain control in that stage. Alternatively, use one of the differential amplifier circuits shown in Chapter 6 which use more than one operational amplifier.

5-6 COMMON MODE REJECTION

Figure 5-4 shows two different situations; Fig. 5-4A shows a single-ended amplifier, while Fig. 5-4B shows a differential amplifier in a similar situation. In these circuits a noise signal, V_n, is placed between the input ground and the output ground. This noise signal might be either AC or DC noise. In Fig. 5-4A we see the case where the noise signal is applied to a single-ended input amplifier. The input signal seen by the amplifier is the algebraic sum of the two independent signals: $V_{in} + V_n$. Because of this fact, the amplifier output signal will see a noise artifact equal to the product of the noise signal amplitude and the amplifier gain, $-A_v V_n$. Now consider the situation of a differential amplifier depicted in Fig. 5-4B. The noise signal in this case is common mode, so is essentially cancelled by the common mode rejection ratio. Of course, in non-ideal amplifiers the actual situation is that the input signal, V_{in}, is subject to the differential gain, while the noise signal, V_n, is subject to the common mode gain. If an amplifier has a CMRR of 90 dB, for example, the gain seen by the noise signal will be 90 dB down from the differential gain.

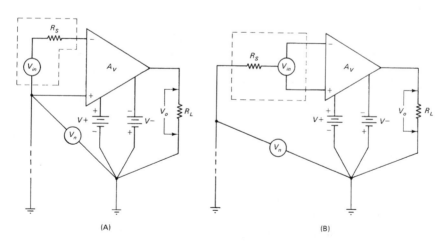

(A)   (B)

FIGURE 5-4 Noise and input signals applied to: (A) single-ended op-amp circuit; (B) differential op-amp circuit.

EXAMPLE 5-3

A DC differential amplifier has a common mode rejection ratio of 100 dB. Calculate the output potential caused by a common mode signal of 1 Vdc.

Solution

$$100 \text{ dB} = 20 \log \left[\frac{V_{cm}}{V_o} \right] = 20 \log \left[\frac{1 \text{ Vdc}}{V_o} \right]$$

Solving for V_o:

$$\frac{100 \text{ dB}}{20} = \log \left[\frac{1 \text{ Vdc}}{V_o} \right]$$

$$5 \text{ dB} = \log \left[\frac{1 \text{ Vdc}}{V_o} \right]$$

$$10^5 \text{ dB} = \frac{1 \text{ Vdc}}{V_o}$$

$$V_o = \frac{1 \text{ Vdc}}{10^5} = 0.00001 \text{ Vdc} \qquad \blacksquare$$

The common mode rejection ratio (CMRR) of the operational amplifier DC differential circuit is dependent principally upon two factors. First, the natural CMRR of the operational amplifier used as the active device. Second, the balance of the resistors, $R1 = R2$ and $R3 = R4$. Unfortunately, the balance is typically difficult to obtain with fixed resistors. We can use a circuit such as Fig. 5-5. In this circuit, $R1$ through $R3$ are exactly the same as in previous circuits. The fourth resistor, however, is a potentiometer (see $R4$). The potentiometer will 'adjust out' the CMRR errors caused by resistor and related mismatches.

A version of the circuit with greater resolution is shown in the inset to Fig. 5-5. In this version the single potentiometer is replaced by a fixed resistor and a potentiometer in series, the sum resistance ($R4A + R4B$) is equal to approximately 20% more than the normal value of $R3$. Ordinarily, the maximum value of the potentiometer is 10–20% of the overall resistance.

The adjustment procedure for either version of Fig. 5-5 is the same (see Fig. 5-6):

1. Connect a zero-center DC voltmeter to the output terminal ($M1$ in Fig. 5-6).
2. Short together inputs A and B, and then connect them to either a signal voltage source, or ground.
3. Adjust potentiometer $R4$ (CMRR ADJ) for zero volts output.
4. If the output indicator (meter $M1$) has several ranges, then switch to a lower range and repeat stages 1–3 above until no further improvement is possible.

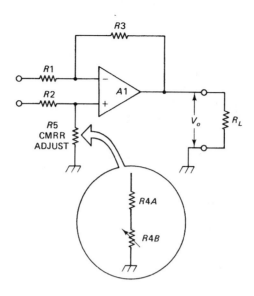

FIGURE 5-5   Making a common mode rejection ratio adjustment control.

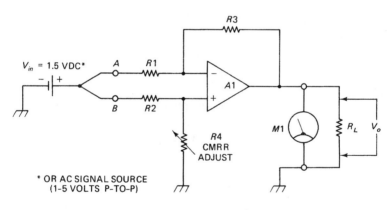

FIGURE 5-6   Test circuit for adjusting CMRR.

Alternatively, connect the output to an audio voltmeter or oscilloscope, and connect the input to a 1 volt to 5 volt peak-to-peak AC signal that is within the frequency range of the particular amplifier. For audio amplifiers, a 400 to 1000 Hz 1 volt signal is typically used.

5-7   PRACTICAL DIFFERENTIAL AMPLIFIER CIRCUITS

Figure 5-7 shows the circuit of the simple DC differential amplifier based on the operational amplifier device. The gain of the circuit is set by the ratio of

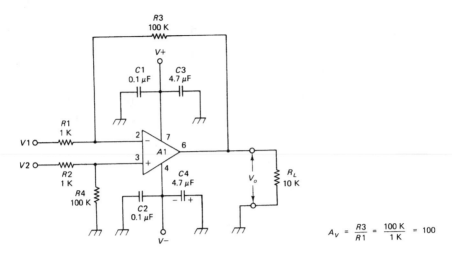

$$A_v = \frac{R3}{R1} = \frac{100\ K}{1\ K} = 100$$

FIGURE 5-7   Practical gain-of-100 DC differential amplifier.

two resistors:

$$A_v = \frac{R3}{R1} \tag{5-29}$$

or

$$A_v = \frac{R4}{R2} \tag{5-30}$$

provided that $R1 = R2$ and $R3 = R4$.

The output voltage of the DC differential amplifier is given by:

$$V_o = \frac{R3}{R1}(V2 - V1) \tag{5-31}$$

or

$$V_o = A_v(V2 - V1) \tag{5-32}$$

where:

V_o is the output voltage
$R3$ is the feedback resistor
$R1$ is the input resistor
A_v is the differential gain
$V1$ is the signal voltage applied to inverting input
$V2$ is the signal voltage applied to the noninverting input

Again, the constraint is that $R1 = R2$, and $R3 = R4$. These two balances must be maintained or the common mode rejection ratio (CMRR) will deteriorate rapidly. In many common applications, the CMRR can be maintained within reason by specifying 5% tolerance resistors for $R1$ through $R4$. But where superior CMRR is required, especially where the differential voltage gain is high, closer tolerance resistors (1% or better) are required.

DESIGN EXAMPLE 5-1

Design a DC differential amplifier with a gain of 100. Assume that the source impedance of the preceding stage is about 100 ohms (refer to Fig. 5-7).

Solution

1. Because the source impedance is 100 ohms, we need to make the input resistors of the DC differential amplifier ten times larger (or more). Thus, the input resistors ($R1$ and $R2$) must be at least 1000 ohms: $R1 = R2 = 1000$ ohms.

2. Set the value of the two feedback resistors (keep in mind that $R3 = R4$). The value of these resistors is found from rewriting Eq. (5-29):

$$R3 = A_v R1$$

$$R3 = (100)(1000)$$

$$R3 = 100\,000 \text{ ohms} \qquad \blacksquare$$

Figure 5-7 shows the finished circuit of this amplifier. The values of the resistors are:

$$R1 = 1 \text{ kohm}$$

$$R2 = 1 \text{ kohm}$$

$$R3 = 100 \text{ kohms}$$

$$R4 = 100 \text{ kohms}$$

For best results, make $R1$, $R2$, $R3$ and $R4$ 1% precision resistors.

The pinouts shown in Fig. 5-7 are for the industry standard 741-family of operational amplifiers. These same pinouts are found on many other different op-amp products as well.

The DC power supply voltages are usually either ±12 volts DC, or ±15 volts DC. Lower potentials can be accommodated, however, if a corresponding reduction in the output voltage swing is tolerable. A typical lower grade op-amp will produce a maximum output voltage that is approximately 3 volts lower than the supply voltage. For example, when $V+$ is 12 volts DC the maximum positive output voltage permitted is $(12 - 3)$, or 9 volts.

The decoupling bypass capacitors shown in Fig. 5-7 are used to keep the circuit stable, especially in those cases where the same DC power supplies are used for several stages. The low value 0.1 μF capacitors ($C1$ and $C2$) are used to decouple high frequency signals. These capacitors should be physically placed as close as possible to the body of the operational amplifier. The high value capacitors ($C3$ and $C4$) are needed to decouple low frequency signals. The reason why we need two values of capacitors is that the high value

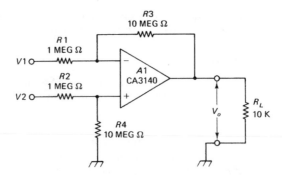

FIGURE 5-8 Gain-of-10 DC differential input amplifier with high input impedance.

capacitors needed for low frequency decoupling are typically electrolytics (tantalum or aluminum), some of which are ineffective at high frequencies. Thus, we must provide smaller value capacitors that (1) have a low enough capacitive reactance to do the job, and (2) are effective at high frequencies.

A different situation is shown in Fig. 5-8. Most differential amplifiers have relatively low input impedances, which is a function of factors such as input bias currents and so forth. The amplifier in Fig. 5-8 uses a high input impedance by virtue of the high values of resistors $R1$ and $R2$. In order to attain this input impedance, however, we need to specify an operational amplifier for $A1$ that has a very low input bias current (i.e. a very high natural input impedance). The BiMOS devices, which use MOSFET input transistors, and the BiFET, which uses JFET input transistors, are good selections for DC differential amplifier circuits when very high input impedances are needed.

5-8 ADDITIONAL DIFFERENTIAL AMPLIFIERS

The simple DC differential amplifier circuits shown in this chapter are useful for low gain applications, and for those applications where a low to moderate input impedance is permissible (e.g. 300 to 200 000 ohms). Where a higher gain is required, then we must resort to a more complex circuit called the operational amplifier instrumentation amplifier, or IA. In the next chapter we will examine the classical three-device IA circuit, as well as several integrated circuit instrumentation amplifiers (ICIA) that offer the advantages of the IA in a single small IC package. Some of those devices are now among the most commonly used in many instrumentation applications, and will be considered in Chapter 6.

5-9 SUMMARY

1. A differential voltage amplifier is a circuit that produces an output voltage signal that is equal to the product of the differential voltage gain and the difference between the potentials applied to the $-IN$ and $+IN$ inputs.

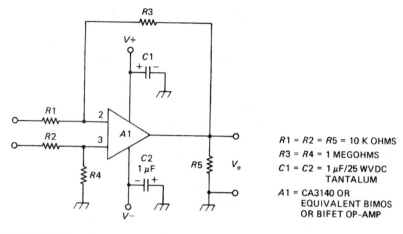

FIGURE 5-9   DC differential amplifier.

$R1 = R2 = R5 = 10 \text{ K OHMS}$
$R3 = R4 = 1 \text{ MEGOHMS}$
$C1 = C2 = 1\,\mu\text{F}/25\text{ WVDC}$
   TANTALUM
$A1 = \text{CA3140 OR}$
   EQUIVALENT BIMOS
   OR BIFET OP-AMP

2. A single op-amp differential amplifier such as Fig. 5-9 has a gain equal to $R3/R1$ or $R4/R2$, assuming that $R1 = R2$ and $R3 = R4$.

3. The common mode gain of a differential amplifier is the gain to a signal that is applied to both −IN and +IN inputs simultaneously. The common mode rejection ratio (CMRR) is the ratio of differential gain to common mode gain.

4. The input impedance of the simple DC differential amplifier (such as Fig. 5-9) is limited to the sum of the two input resistors.

5-10   RECAPITULATION

Now return to the objectives and Pre-quiz questions at the beginning of the chapter and see how well you can answer them. If you cannot answer certain questions, place a check mark to each and review the appropriate parts of the text. Next, try to answer the questions and work the problems below, using the same procedure.

5-11   STUDENT EXERCISES

1. Construct a DC differential amplifier using the circuit of Fig. 5-8. The gain is 100. Short both inputs together, and connect this junction to the output of an audio signal generator. Apply a 3 volt peak-to-peak sinewave signal (frequency between 100 and 1000 Hz). Measure the output signal amplitude using an oscilloscope.

2. In the exercise above, change resistor R4 to 100 kohms, and repeat the experiment.

3. In the first exercise above, replace R4 with a series combination of a 1 megohm potentiometer and a 820 kohm fixed resistor. Using the procedure similar to that in the text adjust for minimum output signal.

4. In all three cases, calculate the common mode rejection ratio, and compare them with each other.

5-12 QUESTIONS AND PROBLEMS ―――――――――――

1. Define differential voltage gain.

2. Define common mode voltage gain.

3. What is the gain of a DC differential amplifier (Fig. 5-9) in which $R1 = R2 = 100$ kohms, and $R3 = R4 = 1$ megohm.

4. What is the gain of a DC differential amplifier (Fig. 5-9) in which $R1 = R2 = 100$ kohms, and $R3 = R4 = 10$ kohms.

5. What is the gain of a DC differential amplifier (Fig. 5-9) in which $R1 = R2 = 500$ ohms, and $R3 = R4 = 1$ kohms.

6. Draw a circuit for a DC differential amplifier in which a fine CMRR ADJUST control covers the range of approximately 80 to 120% of the value of the feedback resistor. Assume a gain of 100 and an input resistance of 1000 ohms.

7. A DC differential amplifier has a differential voltage gain of 120, and a common mode gain of 0.001. Calculate the CMRR in decibels.

8. In the amplifier in the previous question, assume that $V1 = 3.04$ and $V2 = 3.06$. Calculate the output voltage.

9. A 3 Vdc level is applied to a DC differential amplifier as a common mode signal. If the differential gain of the circuit is 200, and the CMRR is 90 dB, what is the output voltage? Assume that no differential signal is present.

10. A DC differential amplifier must be designed with an input impedance of 1 megohm, and a gain of 10. Using only one operational amplifier, design a suitable circuit.

11. A DC differential amplifier must be designed with an input impedance of 1 kohm, and a gain of 1000. Using only one operational amplifier, design a suitable circuit.

12. A DC differential amplifier must be designed with an input impedance of 100 kohms, and a gain of 50. Using only one operational amplifier, design a suitable circuit.

13. A DC differential amplifier must be designed with an input impedance of 10 kohms, and a gain of 200. Using only one operational amplifier, design a suitable circuit.

14. What is the gain of a DC differential amplifier (Fig. 5-9) in which $R1 = R2 = 10$ kohms, and $R3 = R4 = 10$ kohms.

15. What is the gain of a DC differential amplifier (Fig. 5-9) in which $R1 = R2 = 100$ kohms, and $R3 = R4 = 150$ kohms.

16. What is the gain of a DC differential amplifier (Fig. 5-9) in which $R1 = R2 = 470$ kohms, and $R3 = R4 = 1$ megohms.

17. An operational amplifier has a differential voltage gain of 300 000 in the open-loop configuration. If the CMRR is 120 dB, find the common mode voltage gain.

18. Calculate the CMRR in decibels if the common mode voltage gain is 0.5 and the differential voltage gain is 220 000.

19. Calculate the CMRR in decibels if the common mode voltage gain is 0.02 and the differential gain is 150 000.

20. A differential input operational amplifier has a measured CMRR of 80 dB, and a differential voltage gain of 100 dB. Calculate the output voltage that would occur if a common mode voltage of 1 volt is applied to both inputs simultaneously.

21. A common mode voltage of 10 volts DC results in an output voltage of 100 mV, while a 10 mV differential signal voltage results in a 2.5 volt output. Calculate the common mode rejection ratio (CMRR) in decibels.

22. Sketch the circuit for a single op-amp differential amplifier that has a means for adjusting the CMRR. Set the differential voltage gain at 100 and the input impedance to at least 10 kohms. Select values for the resistors used in this circuit.

23. Draw the circuit for a differential DC amplifier with a gain of 1000 and a minimum input impedance of 2200 ohms.

24. A DC differential amplifier based on an operational amplifier has a common mode rejection ratio of 120 dB. Calculate the differential voltage gain if the common mode gain is 0.001.

25. Draw the circuit for a DC differential amplifier that has a gain of 500 and an input impedance of more than 50 megohms. More than one active device will be required in order to avoid using resistors of excessive value.

CHAPTER 6

Instrumentation amplifiers

OBJECTIVES

1. Know the basic properties of the instrumentation amplifier.
2. Know how to design instrumentation amplifier circuits.
3. Understand the range of applications for, and uses of instrumentation amplifiers.
4. Become familiar with various forms of IC instrumentation amplifier (ICIA) devices.

6-1 PRE-QUIZ

These questions test your prior knowledge of the material in this chapter. Try answering them before you read the chapter. Look for the answers (especially those you answered incorrectly) as you read the text. After you have finished studying the chapter try answering these questions again, and those at the end of the chapter (see Section 6-13).

1. List three advantages of the IC instrumentation amplifier.
2. Write the transfer equation for the classical three-device instrumentation amplifier.
3. Why are guard shield connections sometimes used in instrumentation amplifier circuits?
4. Calculate the gain of an INA-101 ICIA circuit in which the gain setting resistor (R_g) is 500 ohms.

6-2 INTRODUCTION

The simple DC differential amplifier discussed in Chapter 5 suffers from several drawbacks. First, there is a limit to the practical input impedance that can be achieved; Z_{in} is approximately equal to the sum of the two input resistors. Second, there is a practical limitation on the gain available from the simple single-device DC differential amplifier. If we attempt to obtain too high gain with ordinary op-amps, then we find that either the input bias current tends to cause large output offset voltages, or the input impedance becomes too low. In this chapter we will demonstrate a solution to these problems in the form of the op-amp *instrumentation amplifier* (IA) circuit. All of these amplifiers are differential amplifiers, but offer superior performance over the simple DC differential amplifiers of the last chapter. The instrumentation amplifier can offer higher input impedance, higher gain, and better common mode rejection than the single-device DC differential amplifier.

6-3 SIMPLE IA CIRCUIT

The simplest form of instrumentation amplifier circuit is shown in Fig. 6-1. In this circuit the input impedance is improved by connecting inputs of a simple DC differential amplifier ($A3$) to two input amplifiers ($A1$ and $A2$) that are each of the unity gain, noninverting follower configuration (their use here is as buffer amplifiers). The input amplifiers offer an extremely large input impedance (a result of the noninverting configuration) while driving the input resistors of the actual amplifying stage ($A3$). The overall gain of this circuit is the same as for any simple DC differential amplifier:

$$A_V = \frac{R3}{R1} \tag{6-1}$$

where A_V is the voltage gain; $R1$ and $R3$ are the resistances of resistors $R1$ and $R3$. Equation (6-1) holds true if $R1 = R2$ and $R3 = R4$.

It is considered best practice if $A1$ and $A2$ are identical operational amplifiers. In fact, it is advisable to use a dual operational amplifier for both $A1$

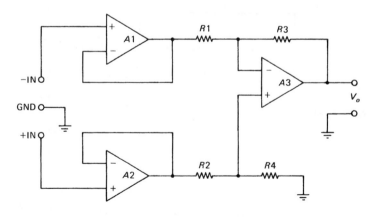

FIGURE 6-1   Simple instrumentation amplifier circuit.

and $A2$. The common thermal environment of the dual amplifier will reduce thermal drift problems. The very high input impedance of superbeta (i.e. Darlington configuration), BiMOS and BiFET operational amplifiers make them ideal for use as the input amplifiers in this type of circuit.

One of the problems with the circuit of Fig. 6-1 is that it wastes two good operational amplifiers. The most common form of instrumentation amplifier circuit uses the input amplifiers to provide voltage gain in addition to a higher input impedance. Such amplifier circuits are discussed in the next section.

6-4   STANDARD INSTRUMENTATION AMPLIFIERS _____

The standard instrumentation amplifier (IA) is shown in Fig. 6-2A. Like the simple circuit discussed above, this circuit uses three operational amplifiers.

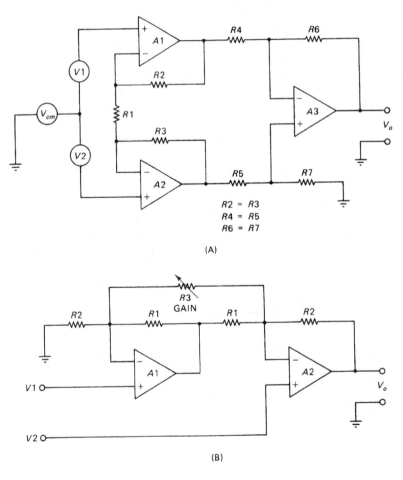

FIGURE 6-2   (A) Standard three op-amp instrumentation amplifier circuit; (B) two op-amp circuit is useful but less desirable.

The biggest difference is that the input amplifiers ($A1$ and $A2$) are used in the 'noninverting follower with gain' configuration. Like the circuit of Fig. 6-1, the input amplifiers are ideally BiMOS, BiFET or superbeta input types for maximum input impedance. Again, for best thermal performance use a dual, triple or quad operational amplifier for this application. The signal voltages shown in Fig. 6-2A follow the standard pattern: voltages $V1$ and $V2$ form the differential input signal ($V2 - V1$), while voltage V_{cm} represents the common mode signal because it affects both inputs equally.

Let's evaluate Fig. 6-2A by first examining the behavior of the input stages, $A1$ and $A2$. In the partial circuit of Fig. 6-3 the output voltage V_c is the difference between the output potential of $A1$ (i.e. $V3$) and the output potential of $A2$ (i.e. $V4$). Given that resistor $R1$ is shared by both $A1$ and $A2$, we count its value as $R1/2$ for each calculation. Our method is to calculate $V3$ when $V2 = 0$, and $V4$ when $V1 = 0$, and then superimpose the result to find $V_c = V4 - V3$.

1.
$$V4 = V2 \left[\frac{R3}{R1/2} + 1 \right] \tag{6-2}$$

2.
$$V3 = V1 \left[\frac{R2}{R1/2} + 1 \right] \tag{6-3}$$

3. Therefore,

$$V4 - V3 = V2 \left[\frac{R3}{R1/2} + 1 \right] - V1 \left[\frac{R2}{R1/2} + 1 \right] \tag{6-4}$$

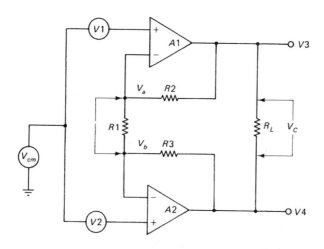

FIGURE 6-3   Input stages for the three op-amp instrumentation amplifier.

4. Because $V_c = V4 - V3$, and $R2 = R3 = R$, we may rewrite Eq. (6-4) in the form:

$$V_c = V2 \left[\frac{R}{R1/2} + 1 \right] - V1 \left[\frac{R}{R1/2} + 1 \right] \tag{6-5}$$

5.
$$V_c = (V2 - V1) \left[\frac{R}{R1/2} + 1 \right] \tag{6-6}$$

6.
$$V_c = (V2 - V1) \left[\frac{R2}{R1} + 1 \right] \tag{6-7}$$

7. Or, in a somewhat better form that identifies specific circuit components:

$$V_c = (V2 - V1) \left[\frac{2\,R2}{R1} + 1 \right] \tag{6-8}$$

In the form of a standard differential amplifier:

$$V_c = V_d A_{V12} \tag{6-9}$$

So, we may conclude by comparing Eqs (6-8) and (6-9):

$$V_d = V2 - V1 \tag{6-10}$$

$$A_{V12} = \frac{2\,R2}{R1} + 1 \tag{6-11}$$

From our discussions in Chapter 5 we know that the gain of $A3$ in Fig. 6-2A is:

$$A_{V3} = \frac{R6}{R4} \tag{6-12}$$

The overall gain of Fig. 6-2A is:

$$A_{V13} = A_{V12} \times A_{V3} \tag{6-13}$$

By substituting Eqs (6-11) and (6-12) into Eq. (6-13) we arrive at the standard transfer equation for the instrumentation amplifier:

$$A_V = \left[\frac{2\,R2}{R1} + 1 \right] \left[\frac{R6}{R4} \right] \tag{6-14}$$

provided that $R2 = R3$, $R4 = R5$ and $R6 = R7$.

It is interesting to note that the condition $R2 \neq R3$ only marginally affects gain, but profoundly affects common mode rejection ratio (CMRR).

EXAMPLE 6-1

Find the gain of the instrumentation amplifier of Fig. 6-2A if the following values of resistors are used: $R1 = 220$ ohms, $R2 = R3 = 2200$ ohms, $R4 = R5 = 10$ kohms, and $R6 = R7 = 82$ kohms.

Solution

$$A_V = \left[\frac{2\,R2}{R1} + 1\right]\left[\frac{R6}{R4}\right]$$

$$A_V = \left[\frac{(2)(2200)}{220} + 1\right]\left[\frac{82\,000}{10\,000}\right]$$

$$A_V = \left[\frac{(4400)}{220} + 1\right]\left[\frac{82\,000}{10\,000}\right]$$

$$A_V = (20 + 1)(8.2) = (21)(8.2) = 172 \qquad \blacksquare$$

An alternative instrumentation amplifier circuit is shown in Fig. 6-2B. This circuit also offers the advantage of high input impedance, but uses only two operational amplifier devices rather than three (as were used in Fig. 6-2A). The gain of this circuit is given by:

$$A_{Vd} = \frac{R2(2\,R1 + R3)}{R1R3} + 1 \qquad (6\text{-}15)$$

In most practical situations the problem will be to select a value for the gain-ranging resistor that is consistent with the required differential voltage gain and the values of the remaining resistors ($R1$ and $R2$). For this application we use the following equation:

$$R3 = \frac{2\,R2}{A_V - 1 - (R2/R1)} \qquad (6\text{-}16)$$

A practical design problem with the simple DC differential amplifier of Chapter 5 is gain control. That problem is, as you will see in Section 6-5, easily solved in instrumentation amplifier circuits.

6-5   GAIN CONTROL FOR THE IA

It is difficult to provide a gain control for a simple DC differential amplifier without adding an extra amplifier stage (for example an inverting follower with a gain of 0 to -1). For the instrumentation amplifier, however, resistor $R1$ can be used as a gain control provided that the resistance does not go to a value near zero ohms (in which case, gain would try to go to infinity). Figure 6-4 shows a revised circuit with resistor $R1$ replaced by a series circuit consisting of fixed resistor $R1A$ and potentiometer $R1B$. This two-resistor circuit prevents the gain from rising to above the level set by $R1A$. Don't use a potentiometer alone in this circuit because it can have disastrous effect

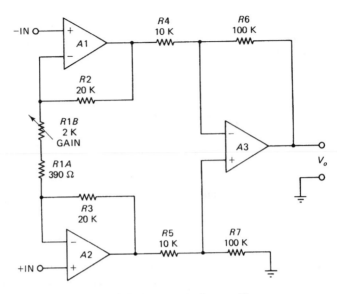

FIGURE 6-4   Variable gain control for instrumentation amplifier.

on the gain. Note in Eq. (6-14) that the term $R1$ appears in the denominator. If the value of $R1$ gets close to zero, then the gain goes very high (in fact, supposedly to infinity if $R1 = 0$). The maximum gain of the circuit by using the fixed resistor in series with the potentiometer is the gain set by the fixed resistor value alone. The gain of the circuit in Fig. 6-4 varies from a minimum of 167 (when $R1B$ is set to 2000 ohms) to a maximum of 1025 (when $R1B$ is zero). The gain expression for Fig. 6-4 is:

$$A_V = \left[\frac{2\,R2}{R1A + R1B} + 1 \right] \left[\frac{R6}{R4} \right] \tag{6-17}$$

or, rewriting Eq. (6-17) to take into account that $R1A$ is fixed,

$$A_V = \left[\frac{2\,R2}{390 + R1B} \right] \left[\frac{R6}{R4} \right] \tag{6-18}$$

where $R1B$ varies from 0 to 2000 ohms.

6-6   COMMON MODE REJECTION RATIO ADJUSTMENT

The instrumentation amplifier is no different from any other practical DC differential amplifier in that there will be imperfect balance for common mode signals. The operational amplifiers are not ideally matched, so there will be a gain imbalance. This gain imbalance is further deteriorated by the inevitable mismatch of the balanced resistor pairs. The result is that the instrumentation amplifier will respond at least to some extent to common

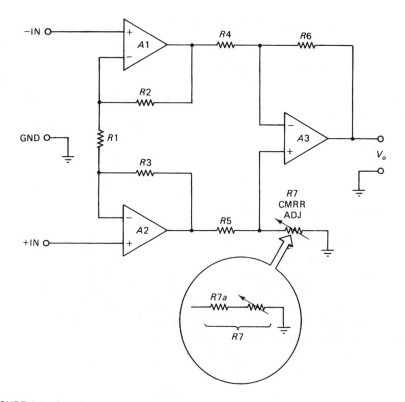

FIGURE 6-5   Common mode rejection ratio (CMRR) adjustment control for instrumentation amplifier.

mode signals. As in the simple DC differential amplifier we can provide a common mode rejection ratio adjustment (CMRR ADJ) by making resistor $R7$ variable (see Fig. 6-5).

One configuration of Fig. 6-5 uses a single potentiometer ($R7$) that has a value that is 10–20% larger than the required resistance of $R6$. For example, if $R6$ is 100 kohms, then $R7$ should be 110 to 120 kohms. Unfortunately, these values are somewhat difficult to obtain, so we would pick a standard value for $R7$ (e.g. 100 kohms), and then select a value for $R6$ that is somewhat lower (e.g. 82 kohms or 91 kohms).

The second configuration of Fig. 6-5 uses a fixed resistor in series with a potentiometer. The general rule is to make $R7A$ approximately 80% of the total required value, and $R7B$ 40% of the required value. As was true in the other configuration, the sum of $R7A$ and $R7B$ is approximately 110 to 120% of the value of resistor $R6$. Resistance values of this sort permit the total resistance to vary from less than, to greater than, the nominally required value. The adjustment of the CMRR ADJUST control follows exactly the same procedure as is given in Chapter 5 for all differential amplifiers.

6-7 AC INSTRUMENTATION AMPLIFIERS

What is the principal difference between 'DC amplifiers' and 'AC amplifiers'? A DC amplifier will amplify both AC and DC signals up to the frequency limit of the particular circuit being used. The AC amplifier, on the other hand, will not pass or amplify DC signals. In fact, AC amplifiers will not pass AC signals of frequencies from close to DC to some lower −3 dB bandpass limit. Gain does not drop to zero below the −3 dB point, but rather falls off at a rate determined by the nature of the circuit. The gain in the region between near-DC and the 'full-gain' frequencies within the passband rises at a rate determined by the design, usually +6 dB/octave (an octave is a 2:1 frequency change). The standard low-end point in the *frequency response curve* is defined as the frequency at which the gain drops off −3 dB from the full gain.

Figure 6-6A shows a version of the instrumentation amplifier circuit that is specially designed as an AC amplifier. The input circuitry of $A1$ and $A2$ is modified by placing a capacitor in series with each op-amp's noninverting input. Resistors $R8$ and $R9$ are used to keep the input bias currents of $A1$ and $A2$ from charging capacitors $C1$ and $C2$. In some modern low input current operational amplifiers these resistors are optional because of the extremely low levels of bias current that are normally present.

The −3 dB frequency of the amplifier in Fig. 6-6A is a function of the input capacitors and resistors (assuming that $R8 = R9 = R$ and $C1 = C2 = C$):

$$F_{Hz} = \frac{1\,000\,000}{2\pi RC_{\mu F}} \qquad (6\text{-}19)$$

where:

F_{Hz} is the −3 dB frequency in hertz (Hz)
R is the resistance in ohms (Ω)
$C_{\mu F}$ is the capacitance C in microfarads (μF)

EXAMPLE 6-2

Find the lower −3 dB breakpoint frequency if $R9 = R10 = 10$ megohms, and $C1 = C2 = 0.1$ μF.

Solution

$$F_{Hz} = \frac{1\,000\,000}{2\pi RC_{\mu F}}$$

$$F_{Hz} = \frac{1\,000\,000}{(2)(3.14)(10^7\ \Omega)(0.1\ \mu F)} = 0.159\ \text{Hz} \qquad \blacksquare$$

The equation given above for frequency response is not the most useful form. In most practical cases you will know the required frequency response from evaluation of the application. Furthermore, you will know the value

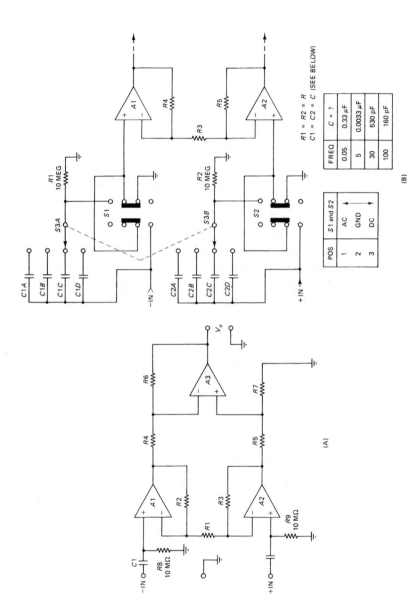

S1 and S2	
POS	
1	AC
2	GND ↔
3	DC

FREQ	C = ?
0.05	0.33 μF
5	0.0033 μF
30	530 pF
100	160 pF

R1 = R2 = R
C1 = C2 = C (SEE BELOW)

(B)

(A)

FIGURE 6-6   (A) AC-coupled input instrumentation amplifier; (B) frequency response tailoring of the input circuit and DC-GND-AC switching.

of the input resistors ($R9$ and $R10$) because they are selected for high input impedance, and are by convention either $> 10 \times$ or $> 100 \times$ the source impedance (depending upon the application). Typically, these resistors are selected to be 10 megohms. You will therefore need to select the capacitor values from Eq. (6-20) below:

$$C_{\mu F} = \frac{1\,000\,000}{2\pi RF} \qquad (6\text{-}20)$$

where:

$C_{\mu F}$ is the capacitance of $C1$ and $C2$, in microfarads (μF)

R is the resistance of $R9$ and $R10$ in ohms (Ω)

F is the -3 dB frequency in hertz (Hz)

The AC instrumentation amplifier can be adapted to all the other modifications of the basic circuit discussed earlier in this chapter. We may, for example, use a gain control (replace $R1$ with a fixed resistor and a potentiometer), or add a CMRR ADJ control. In fact, these adaptations are probably necessary in most practical AC IA circuits.

In many instrumentation amplifier applications it is desirable to provide selectable AC or DC coupling, as well as the ability to ground the input of the amplifiers. This latter feature is especially desirable in circuits where an oscilloscope, strip-chart paper recorder, digital data logger or computer is used to receive the data. By grounding the input of the amplifier (without also grounding the source, which could be dangerous), it is possible to set (or at least determine) the $V_d = 0$ baseline. Figure 6-6B shows a modified input circuit that uses a switch to select AC-GND-DC coupling. In addition, a second switch is provided that sets the low-end -3 dB frequency response point according to Eq. (6-20) above. A table of popular frequency response limits is shown inset to Fig. 6-6B.

6-7.1 Design example: an electrocardiograph (ECG) amplifier

The heart in man and animals produces a small electrical signal that can be recorded through skin surface electrodes and displayed on an oscilloscope or paper strip-chart recorder. This signal is called the *electrocardiograph* or *ECG* signal. The peak values of the ECG signal are on the order of one millivolt (1 mV). In order to produce a 1 volt signal to apply to a recorder or oscilloscope, then we need a gain of 1000 mV/1 mV, or 1000. Our ECG amplifier, therefore, must provide a gain of 1000 or more. Furthermore, because skin has a relatively high electrical resistance (1 to 20 kohms), the ECG amplifier must have a very high input impedance.

Another requirement for the ECG amplifier is that it be an AC amplifier. The reason for this requirement is that metallic electrodes applied to the electrolytic skin produces a *halfcell potential*. This potential tends to be on the order of 1 to 2 volts, so it is more than 1000 times higher than the signal voltage. By making the amplifier respond only to AC, we eliminate the artifact caused by the DC halfcell potential.

The frequency selected for the −3 dB point of the ECG amplifier must be very low, close to DC, because the standard ECG waveform contains some very low frequency components. The typical ECG signal has significant Fourier frequency components in the range 0.05 to 100 Hz, which is the industry standard frequency response for diagnostic ECG amplifiers (some clinical monitoring ECG amplifiers use 0.05 Hz to 40 Hz to eliminate muscle artifact due to patient movements).

The typical ECG amplifier has differential inputs because the most useful ECG signals are differential in nature, and in order to suppress 50/60 Hz hum picked-up on the leads and the patient's body. In the most simple case, the right arm (RA) and left arm (LA) electrodes form the inputs to the amplifier, with the right leg (RL) defined as the common (see Fig. 6-7). The basic configuration of the amplifier in Fig. 6-7 is the AC-coupled instrumentation amplifier discussed earlier. The gain for this amplifier is set to slightly more than ×1000, so a 1 mV ECG peak signal will produce a 1 volt output from this amplifier. Because of the high gain it is essential that the amplifier be well balanced. This requirement suggests the use of a dual amplifier for A1 and A2. An example might be the CA-3240 device, which is a dual BiMOS device that is essentially two CA-3140s in a single eight-pin miniDIP package. Also in the interest of balance, 1% or less tolerance resistors should be used for the equal pairs.

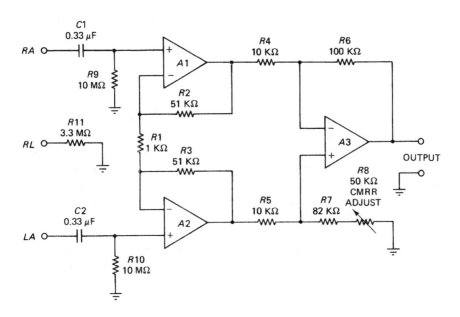

FIGURE 6-7   Simple electrocardiograph amplifier.

The lower end −3 dB frequency response point is set by the input resistors and capacitors. In this case, the combination forms a response of:

$$F_{Hz} = \frac{1\,000\,000}{2\pi RC}$$

$$F_{Hz} = \frac{1\,000\,000}{(2)(3.14)(10^7 \ \Omega)(0.33 \ \mu F)}$$

$$F_{Hz} = \frac{1\,000\,000}{20\,724\,000} = 0.048 \ Hz$$

The CMRR ADJUST control in Fig. 6-7 is usually a ten to twenty-turn trimmer potentiometer. It is adjusted in the following manner:

1. Short together the RA, LA and RL inputs.

2. Connect a DC voltmeter to the output (either a digital voltmeter or an analog meter with a 1.5 volt DC scale (alternatively, a DC-coupled oscilloscope can be used, but be sure to identify the zero baseline).

3. Adjust CMRR ADJUST control (R7) for zero volts output.

4. Disconnect the RL terminal, and connect a signal generator between RL and the still-connected RA/LA terminal. Adjust the output of the signal generator for a sinewave frequency in the range 10 to 40 Hz, and a peak-to-peak potential of 1 volt to 3 volts.

5. Using an AC-scale on the voltmeter (or an oscilloscope), again adjust CMRR ADJUST (R7) for the smallest possible output signal. It may be necessary to readjust the voltmeter or oscilloscope input range control to observe the best null.

6. Remove the RA/LA short. The ECG amplifier is ready to use.

A suitable post-amplifier for the ECG preamplifier is shown in Fig. 6-8. This amplifier is placed in the signal line between the output of Fig. 6-7 and the input of the oscilloscope or paper chart recorder used to display the waveform. The gain of the post-amplifier is variable from 0 to +2; it will produce a 2 volt maximum output when a 1 mV ECG signal provides a 1 volt output from the preamplifier. Because of the high level signals used, this amplifier can use ordinary 741 operational amplifiers.

The frequency response of the amplifier is set to an upper −3 dB point of 100 Hz, with the response dropping off at a −6 dB/octave rate above that frequency. This frequency response point is determined by capacitor C3 operating with resistor R12:

$$F_{Hz} = \frac{1\,000\,000}{2\pi RC}$$

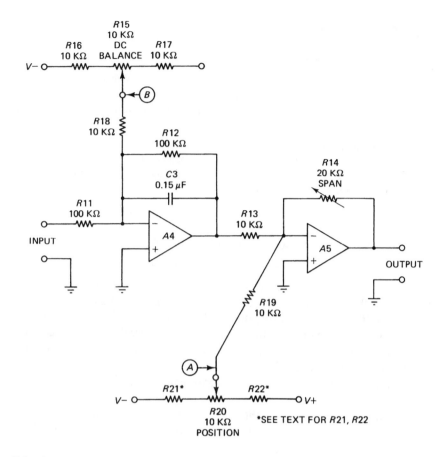

FIGURE 6-8   Post-amplifier circuit.

$$F_{Hz} = \frac{1\,000\,000}{(2)(3.14)(10^5\ \Omega)(0.015\ \mu F)}$$

$$F_{Hz} = \frac{1\,000\,000}{9420}\ Hz = 106\ Hz$$

There are three controls in the post-amplifier circuit: *span, position* and *DC balance.* The span control is the 0 to 2 gain control, and the label 'span' reflects instrumentation users language, rather than electronics language. The position control sets the position of the output waveform on the display device. Resistors $R21$ and $R22$ are selected to limit the travel of the beam or pen to full scale. Set these resistors so that the maximum potential at the end terminals of $R20$ corresponds to full-scale deflection of the display device.

The DC BALANCE control is used to cancel the collective effects of offset potentials created by the various stages of amplification. This control is adjusted as follows:

1. Follow the CMRR ADJUST procedure, and then reconnect the short-circuit at RA, LA and RL. The voltmeter is moved to the output of Fig. 6-8.

2. Adjust the position control for zero volts at point A.

3. Adjust the DC BALANCE control for zero volts at point B.

4. Adjust the SPAN control (R14) through its entire range from zero to maximum while monitoring the output voltage. If the output voltage does not shift, then no further adjustment is needed.

5. If the output voltage in the previous step varied as the span control is varied, then adjust DC balance until varying the span control over its full-range does not produce an output voltage shift. Repeat this step several times until no further improvement is possible.

6. Remove the RA, LA, RL short; the amplifier is ready for use.

6-8 IC INSTRUMENTATION AMPLIFIERS: SOME EXAMPLES

The operational amplifier truly revolutionized analog circuit design. For a long time, the only additional advances were that op-amps became better and better (they became nearer the ideal op-amp of textbooks). While that was an exciting development, they were not really new devices. The next big breakthrough came when the analog device designers made an IC version of the instrumentation amplifier in Fig. 6-2A, the integrated circuit instrumentation amplifier (ICIA). Today, several manufacturers offer substantially improved ICIA devices.

The Burr-Brown INA-101 (Fig. 6-9) is a popular ICIA device; a sample INA-101 circuit is shown in Fig. 6-10. This ICIA amplifier is simple to

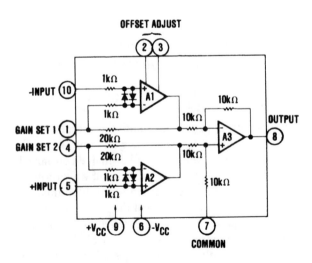

FIGURE 6-9   Commercial integrated circuit instrumentation amplifier (courtesy Burr-Brown).

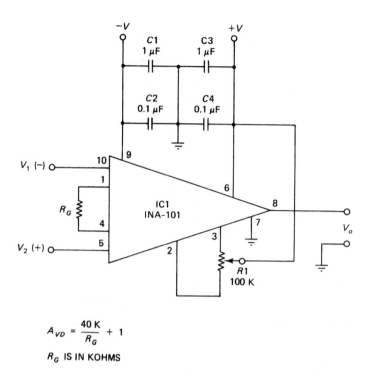

$$A_{VD} = \frac{40 \text{ K}}{R_G} + 1$$

R_G IS IN KOHMS

FIGURE 6-10   Circuit for commercial IC instrumentation amplifier (B-B INA-101).

connect and use. There are only DC power connections, differential input connections, offset adjust connections, ground and an output. The gain of the circuit is set by:

$$A_{vd} = \frac{40 \text{ k}\Omega}{R_g} + 1 \qquad (6\text{-}21)$$

The INA-101 is basically a low-noise, low input bias current integrated circuit version of the IA of Fig. 6-2A. The resistors labeled $R2$ and $R3$ in Fig. 6-2A are internal to the INA-101, and are 20 kohms each, hence the '40 kΩ' term in Eq. (6-21).

Potentiometer $R1$ in Fig. 6-10 is used to null the offset voltages appearing at the output. An offset voltage is a voltage that exists on the output at a time when it should be zero (i.e. when $V1 = V2$, so that $V1 - V2 = 0$). The offset voltage might be internal to the amplifier, or, a component of the input signal. DC offsets in signals are common, especially in biopotentials amplifiers such as ECG and EEG, and chemical transducers such as pH, pO2 and pCO2.

Another ICIA is the LM-363-xx device shown in Figure 6-11; the miniDIP version is shown in Fig. 6-11A (an 8-pin metal can is also available), while a typical circuit is shown in Fig. 6-11B. The LM-363-xx device is a fixed

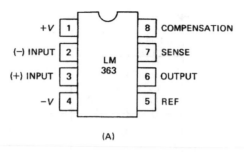

(A)

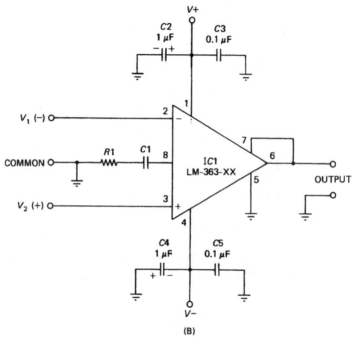

(B)

FIGURE 6-11   LM-363 fixed gain IC instrumentation amplifier: (A) package and pinouts; (B) circuit using the LM-363.

gain ICIA. There are three versions of the LM-363-xx enumerated according to gain:

Designation	Gain (A_V)
LM-363-10	×10
LM-363-100	×100
LM-363-500	×500

The LM-363-xx is useful in places where one of the standard gains is required, and there is minimum space available. Two examples spring to mind. The LM-363-xx can be used as a transducer preamplifier, especially in noisy signal areas; the LM-363-xx can be built onto (or into) the transducer to build up its signal before sending it to the main instrument or signal acquisition computer. Another possible use is in biopotentials amplifiers. Biopotentials are typically very small, especially in lab animals. The LM-363-xx can be mounted on the subject and a higher level signal sent to the main instrument.

A selectable gain version of the LM-363 device is shown in Fig. 6-12A; the 16-pin DIP package is shown in Fig. 6-12A, while a typical circuit is shown in Fig. 6-12B. The type number of this device is LM-363-AD, which distinguishes it from the LM-363-xx devices. The gain can be ×10, ×100 or ×1000 depending upon the programming of the gain setting pins (2, 3 and 4). The programming protocol is as follows:

Gain desired	Jumper pins
×10	(All open)
×100	3 & 4
×1000	2 & 4

Switch $S1$ in Fig. 6-12B is the GAIN SELECT switch. This switch should be mounted close to the IC device, but is quite flexible in mechanical form.

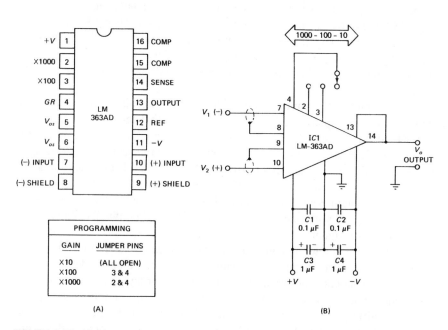

(A)

(B)

FIGURE 6-12 IC instrumentation amplifier with switch selectable gain: (A) package and pinouts; (B) typical circuit.

The switch could also be made from a combination of CMOS electronic switches (e.g. 4066).

The DC power supply terminals are treated in a manner similar to the other amplifiers. Again, the 0.1 μF capacitors need to be mounted as close as possible to the body of the LM-363-AD.

Pins 8 and 9 are guard shield outputs. These pins are a feature that makes the LM-363-AD more useful for many instrumentation problems than other models. By outputting a signal sample back to the shield of the input lines, we can increase the common mode rejection ratio. This feature is used a lot in bipotentials amplifiers and in other applications where a low-level signal must pass through a strong interference (high noise) environment. Guard shield theory will be discussed below.

The LM-363 devices will operate with DC supply voltages of ±5 volts to ±18 volts DC, with a common mode rejection ratio (CMRR) of 130 dB. The 7 nV/$\sqrt{\text{Hz}}$ noise figure makes the device useful for low noise applications (a 0.5 nV model is also available).

6-9 GUARD SHIELDING

One of the properties of the differential amplifier, including instrumentation amplifiers, is that it tends to suppress interfering signals from the environment. The common mode rejection process is at the root of this capability. When an amplifier is used in a situation where it is connected to an external signal source through wires, those wires are subjected to strong local 50/60 Hz AC fields from nearby power line wiring. Fortunately, in the case of the differential amplifier the field affects both lines equally, so the induced interfering signal is cancelled out by the common mode rejection property of the amplifier.

Unfortunately, the cancellation of interfering signals is not total. There may be, for example, imbalances in the circuit that tend to deteriorate the CMRR of the amplifier. These imbalances may be either internal or external to the amplifier circuit. Figure 6-13 shows a common scenario. In this figure we see the differential amplifier connected to shielded leads from the signal source, V_{in}. Shielded lead wires offer some protection from local fields, but there is a problem with the standard wisdom regarding shields: it is possible for shielded cables to manufacture a valid differential signal voltage from a common mode signal!

Figure 6-13B shows an equivalent circuit that demonstrates how a shielded cable pair can create a differential signal from a common mode signal. The cable has capacitance between the center conductor and the shield conductor surrounding it. In addition, input connectors and the amplifier equipment internal wiring also exhibits capacitance. These capacitances are lumped together in the model of Fig. 6-13B as C_{S1} and C_{S2}. As long as the source resistances and shunt resistances are equal, and the two capacitances are

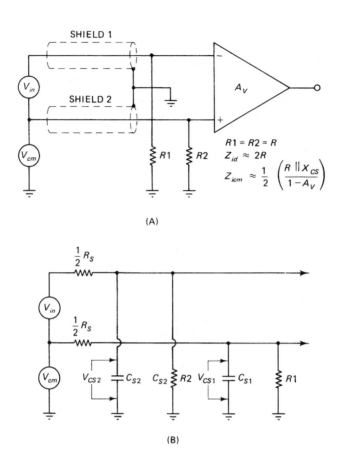

FIGURE 6-13　(A) Input shielding; (B) equivalent circuit.

equal, there is no problem with circuit balance. But inequalities in any of these factors (which are commonplace) creates an unbalanced circuit in which common mode signal V_{cm} can charge one capacitance more than the other. As a result, the difference between the capacitance voltages, V_{CS1} and V_{CS2}, is seen as a valid differential signal.

A low-cost solution to the problem of shield-induced artifact signals is shown in Fig. 6-14A. In this circuit a sample of the two input signals are fed back to the shield, which in this situation is not grounded. Alternatively, the amplifier output signal is used to drive the shield. This type of shield is called a *guard shield*. Either double shields (one on each input line) as shown, or a common shield for the two inputs can be used.

An improved guard shield example for the instrumentation amplifier is shown in Fig. 6-14B. In this case a single shield covers both input lines, but it is possible to use separate shields. In this circuit a sample of the two input signals is taken from the junction of resistors $R8$ and $R9$, and fed to the input of a unity gain buffer/driver 'guard amplifier' (*A4*). The output of *A4* is used to drive the guard shield.

Perhaps the most common approach to guard shielding is the arrangement shown in Fig. 6-14C. Here we see two shields used; the input cabling is double-shielded insulated wire. The guard amplifier drives the inner shield, which serves as the guard shield for the system. The outer shield is grounded at the input end in the normal manner, and serves as an electromagnetic interference suppression shield.

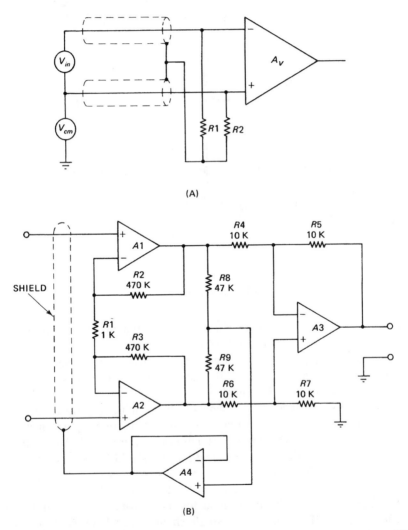

(A)

(B)

FIGURE 6-14 (A) Simple resistor guard shield driver; (B) using a guard shield amplifier; (C) use of guard shield amplifier and double shielded input wiring.

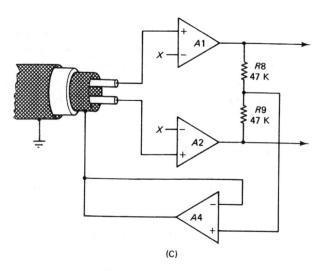

(C)

FIGURE 6-14 (continued)

6-10 SUMMARY

1. The instrumentation amplifier circuit overcomes several difficulties of the simple DC differential amplifier. It provides higher input impedance levels, improved common mode rejection ratio (CMRR) and higher achievable gain levels.

2. The basic instrumentation amplifier uses three operational amplifiers; two input amplifiers (A1 and A2) are connected in the noninverting follower with gain configuration, while the third (A3) is an output amplifier connected as a DC differential amplifier. Amplifiers A1/A2 drive A3.

3. Gain control on the instrumentation amplifier is controllable by adjusting a single 'input' resistor that is common to both input amplifiers.

4. AC-coupled instrumentation amplifiers are formed by using capacitors in series with the inputs. In most circuits each input will require a very high value resistor to common in order to prevent the input capacitors from being charged by the op-amp input bias currents.

6-11 RECAPITULATION

Now return to the objectives and Pre-quiz questions at the beginning of the chapter and see how well you can answer them. If you cannot answer certain questions, place a check mark to each and review the appropriate parts of the text. Next, try to answer the questions and work the problems below, using the same procedure.

6-12 STUDENT EXERCISES

1. Using a circuit such as Fig. 6-2A design and build an instrumentation amplifier with a gain of (a) 100, (b) 500, (c) 1000. Measure the gain and CMRR in each case.

2. Using a circuit such as Fig. 6-6A, design an AC-coupled instrumentation amplifier that has a lower −3 dB point of 10 Hz in its frequency response characteristic. Measure the frequency response from near-DC to 1000 Hz, using at least ten data points between 0.1 Hz and 100 Hz.

3. In the exercise above change the input resistance values and/or the input capacitor values to form frequency responses of 0.1 Hz, 1 Hz, or 100 Hz. Measure the frequency response from near-DC to 1000 Hz.

6-13 QUESTIONS AND PROBLEMS

1. Find the gain of the instrumentation amplifier of Fig. 6-2A if the following values of resistors are used: $R1 = 330$ ohms, $R2 = R3 = 20\,000$ ohms, $R4 = R5 = 10$ kohms, and $R6 = R7 = 56$ kohms.

2. Find the gain of the instrumentation amplifier of Fig. 6-2A if the following values of resistors are used: $R1 = 1$ kohm, $R2 = R3 = 10$ kohms, $R4 = R5 = 10$ kohms, and $R6 = R7 = 100$ kohms.

3. Find the lower −3 dB breakpoint frequency of an AC instrumentation amplifier such as Fig. 6-6A if $R8 = R9 = 10$ megohms, and $C1 = C2 = 0.33$ μF.

4. Find the lower −3 dB breakpoint frequency in an AC instrumentation amplifier such as Fig. 6-6A if $R8 = R9 = 10$ megohms, and $C1 = C2 = 0.01$ μF.

5. Design an instrumentation amplifier for a Wheatstone bridge transducer that will drive an oscilloscope trace full scale (8 cm) when the input selector is set to 1 Vdc/cm. Assume that the full-scale output of the transducer is 10 mV.

6. Design an AC instrumentation amplifier that has a gain of 2000, an input impedance of 10 megohms or more, and a lower end AC frequency response of 0.5 Hz.

7. Design an AC instrumentation amplifier that has a gain of 1 to 1000, an input impedance of 10 megohms or more, and a lower end AC frequency response of 10 Hz.

8. Sketch a circuit for a standard instrumentation amplifier based on three operational amplifiers.

9. List at least three advantages of the instrumentation amplifier over the standard DC differential op-amp circuit.

10. What advantage is gained by making at least the two input op-amps part of the same IC package?

11. Derive the transfer function for the standard three op-amp instrumentation amplifier using the Kirchhoff's law method used in this book. Show your work.

12. Sketch the circuit for an instrumentation amplifier that uses just two operational amplifiers.

13. Calculate the differential gain of a standard three op-amp instrumentation amplifier if all resistors are 10 kohms except the resistor connected between the inverting inputs of the two input amplifiers ($R1$), which is 100 ohms.

14. Calculate the differential voltage gain of a two op-amp instrumentation amplifier if $R1 = 10$ kohms, $R2 = 150$ kohms, and $R3 = 10$ kohms.

15. Select a value for $R3$ in the two op-amp instrumentation amplifier if the other resistors are all 10 kohms and the required voltage gain is 100.

16. Sketch the circuit for a three op-amp instrumentation amplifier that has a fine resolution CMRR ADJ control. Select appropriate resistor values for a fixed voltage gain of 500. Show your calculations to justify your selections.

17. Write out a brief procedure for adjusting CMRR in the amplifier above: (a) when the input signal is 1 volt DC, and (b) when the input signal is a 10 volt peak-to-peak 100 Hz sinewave.

18. Draw the circuit for a bioelectric amplifier that has a lower -3 dB frequency response point of 0.05 Hz or less (but not DC), an upper -3 dB frequency response point of 100 Hz, an input impedance of at least 10 megohms, and a voltage gain that is variable from about 200 to more than 2000.

19. Draw the circuit for an ECG preamplifier that has a gain of 1000 and provides some protection against electrical shock in the common line for the patient.

20. Why must ECG and other bioelectric preamplifiers use AC-coupled input circuitry?

21. Draw the circuit for an instrumentation amplifier input circuit that offers 10 megohms input impedance (or more), with the following switch selectable options: DC coupling, AC coupling, and input to the amplifier (but not signal input connector!) grounded.

22. Draw the circuit for a post-amplifier suitable for use with an ECG preamplifier. Label component values. The gain should be not more than ten, but unity is acceptable.

23. A Burr-Brown INA-101 instrumentation amplifier is used in a circuit that must offer a gain of 1000, and input impedance of 10 megohms or more, and an AC-coupled lower -3 dB point of 0.1 Hz or less. Calculate the values of: (a) the gain-setting resistor, and (b) the coupling capacitors.

24. What is the differential voltage gain of an LM-363-100?

25. Why are guard shield circuits sometimes used in instrumentation amplifier circuits?

26. Draw the circuit for a universal guard shield driver that will work on a three op-amp instrumentation amplifier.

27. Draw a circuit showing the guard shield connections in the LM-363-AD ICIA device.

CHAPTER 7

Isolation amplifiers

OBJECTIVES

1. Understand the types of applications for which isolation amplifiers are required.
2. Learn the different approaches to isolation amplifier design.
3. Examine typical applications for isolation amplifiers.
4. Learn to design simple instrumentation circuits based on isolation amplifiers.

7-1 PRE-QUIZ

These questions test your prior knowledge of the material in this chapter. Try answering them before you read the chapter. Look for the answers (especially those you answered incorrectly) as you read the text. After you have finished studying the chapter try answering these questions again, and those at the end of the chapter (see Section 7-10).

1. List two potential applications for isolation amplifiers, and why each is a candidate.
2. List three different approaches to isolation amplifier design.
3. Under what circumstances is a battery powered amplifier considered 'isolated'?
4. From what are the inputs of an isolation amplifier isolated?

7-2 INTRODUCTION

There are a number of applications in which ordinary solid-state amplifiers are either in danger themselves, or present a danger to the users. An example of

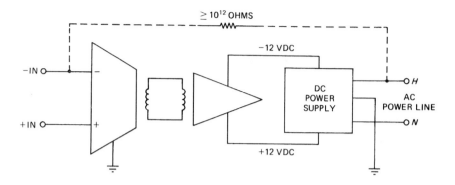

FIGURE 7-1 Block diagram of isolation amplifier.

the former is an amplifier in a high voltage experiment such as a biochemist's electrophoresis system (Section 7-4.3), while the latter is represented by cardiac monitors and other devices used in clinical medicine. Many of the commercial products of this type now available on the market are not, strictly speaking, integrated circuits, but rather are hybrids. Nonetheless, it is important to cover these devices in any book on linear IC amplifiers.

An *isolation amplifier* (Fig. 7-1) has an extremely high impedance between the signal inputs and those power supply terminals that are connected to a DC power supply that are, in turn, connected to the AC power mains. Thus, in isolation amplifiers there is an extremely high resistance (on the order of 10^{12} ohms) between the amplifier input terminals and the AC power line. In the case of medical equipment, the goal is to prevent minute leakage currents from the 50/60 Hz AC power lines from being applied to the patient. Current levels that are normally negligible to humans can theoretically be fatal to a hospital patient in situations where the body is invaded by devices that are electrical conductors. In other cases, the high impedance is used to prevent high voltages at the signal inputs from adversely affecting the rest of the circuitry. Modern isolation amplifiers can provide more than 10^{12} ohms of isolation between the AC power lines and the signal inputs.

Several different circuit symbols are used to denote the isolation amplifier in schematic diagrams, but the one that is the most common is shown in Fig. 7-2. It consists of the regular triangular amplifier symbol broken in the middle to indicate isolation between the A and B sections. The following connections are usually found on the isolation amplifier:

Non-isolated A Side. V+ and *V−* DC power supply lines (to be connected to a DC supply powered by the AC lines), output to the rest of the (non-isolated) circuitry, and (in some designs) a nonisolated ground or common. This ground is connected to the chassis or main system ground also served by the main DC power supplies.

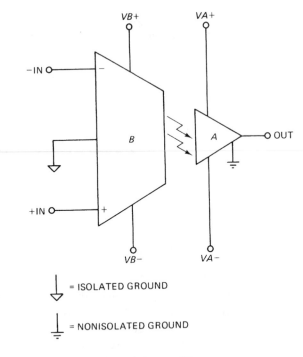

VB+   VA+

−IN

B

A

OUT

+IN

VB−   VA−

= ISOLATED GROUND

= NONISOLATED GROUND

FIGURE 7-2   Schematic symbol for isolation amplifier.

Isolated B Side. Isolated $V+$ and $V-$, isolated ground or common and the signal inputs. The isolated power supply and ground are not connected to the main power supply or ground systems. Batteries are sometimes used for the isolated side, while in other cases special isolated DC power supplies derived from the main supplies are used (of which, more later).

7-3 APPROACHES TO ISOLATION AMPLIFIER DESIGN _____

Different manufacturers use different approaches to the design of isolation amplifiers. Common circuit approaches to isolation include: *battery power, carrier operated, optically coupled,* and *current loading.* These methods are discussed in detail in the sections below.

7-3.1 Battery powered isolation amplifiers

The battery approach to isolation amplifier design is perhaps the simplest to implement, but it is not always most suitable due to problems inherent in battery upkeep. A few products exist, however, that use a battery-powered front-end amplifier, even though the remainder of the equipment is powered from the AC power line. Other products are entirely battery powered. A battery-powered amplifier or instrument is isolated from the AC power mains only if the battery is disconnected from the charging circuit during use. In some battery-powered instruments used in medicine, mechanical interlocks and electrical logic circuitry prevents the instrument from being turned on if the AC power cord is still attached. Later in this chapter we will study a battery-powered cardiac output computer as a design example of this type of isolation.

7-3.2 Carrier-operated isolation amplifiers

Figure 7-3A shows an isolation amplifier that uses the carrier signal technique to provide isolation. The circuitry inside of the dashed line is isolated from the AC power lines (in other words, the B side of Fig. 7-2). The voltage gain of the isolated section is typically in the range ×10 to ×500.

The isolation is provided by separation of the ground, power supply and signal paths into two mutually exclusive sections by high frequency transformers $T1$ and $T2$. These transformers have a design and core material that works very well in the ultrasonic (20 kHz to 500 kHz) region, but is very inefficient at the 50/60 Hz frequency used by the AC power lines. This design feature allows the transformers to easily pass the high frequency carrier signal, while severely attenuating 50/60 Hz energy. Although most models use a carrier frequency in the 50 kHz to 60 kHz range, examples of carrier amplifiers exist over the entire 20 kHz to 500 kHz range.

The carrier oscillator signal is coupled through transformer $T1$ to the isolated stages. Part of the energy from the secondary of $T1$ is directed to the modulator stage; the remainder of the energy is rectified and filtered, and then used as an isolated DC power supply. The DC output of this power supply is used to power the input B amplifiers and the modulator stage.

An analog signal applied to the input is amplified by $A1$, and is then applied to one input of the modulator stage. This stage amplitude modulates the signal onto the carrier. Transformer $T2$ couples the signal to the input of the demodulator stage on the nonisolated side of the circuit. Either envelope or synchronous demodulation may be used, although the latter is considered superior. Part of the demodulator stage is a low-pass filter that removes any residual carrier signal from the output signal. Ordinary DC amplifiers following the demodulator complete the signal processing chain.

An example of a synchronous demodulator circuit is shown in Fig. 7-3B. These circuits are based on switching action. Although the example shown uses bipolar PNP transistors as the electronic switches, other circuits use NPN transistors, FETs or CMOS electronic switches (e.g. 4066 device).

The signal from the modulator has a fixed frequency in the range from 20 kHz to 500 kHz, and is amplitude modulated with the input signal from the isolated amplifier. This signal is applied to the emitters of transistors $Q1$ and $Q2$ (via $T1$) in push-pull. On one half of the cycle, therefore, the emitter of $Q1$ will be positive with respect to the emitter of $Q2$. On alternate

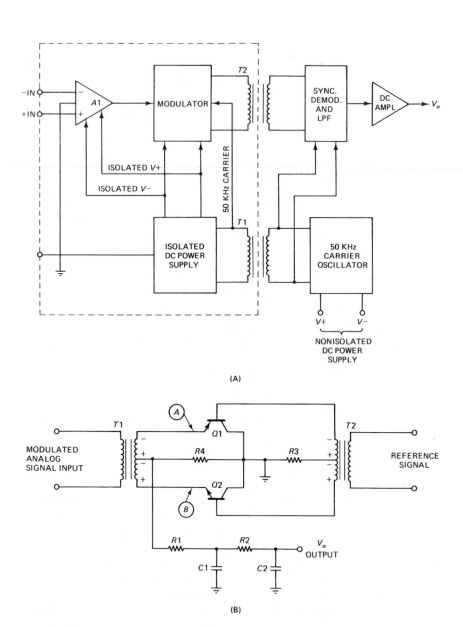

FIGURE 7-3 (A) Carrier-method for achieving amplifier isolation; (B) phase sensitive detector circuit.

half-cycles, the opposite situation occurs: $Q2$ is positive with respect to $Q1$. The bases of $Q1$ and $Q2$ are also driven in push-pull, but by the carrier signal (called here the 'reference signal'). This action causes transistors $Q1$ and $Q2$ to switch on and off out of phase with each other.

On one half of the cycle, the polarities are as shown in Fig. 7-3B; transistor $Q1$ is turned on. In this condition point A on $T1$ is grounded. The voltage developed across load resistor $R4$ is positive with respect to ground.

On the alternate half-cycle, $Q2$ is turned on, so point B is grounded. But the polarities have reversed, so the polarity of the voltage developed across $R4$ is still positive. This causes a full-wave output waveform across $R4$, which when low-pass filtered becomes a DC voltage level proportional to the amplitude of the input signal. This same description of synchronous demodulators also applies to the circuits used in some carrier amplifiers (a specialized laboratory amplifier used for low-level signals).

A variation on this circuit replaces the modulator with a *voltage controlled oscillator* (VCO) that allows the analog signal to *frequency modulate* (FM) a carrier signal generated by the VCO. The power supply carrier signal is still required, however. A phase detector, phase-locked loop (PLL), or pulse-counting FM detector on the nonisolated side recovers the signal.

7-3.3 Optically coupled isolation amplifier circuits

Electronic optocouplers (also called optoisolators) are sometimes used to provide the desired isolation. In early designs of this class, a light emitting diode (LED) was mounted together with a photoresistor or phototransistor. Modern designs, however, use integrated circuit (IC) optoisolators that contain an LED and phototransistor inside of a single DIP IC package.

There are actually several approaches to optical coupling. Two common methods are the *carrier* and *direct methods*. The carrier method is the same as discussed in the previous section, except that an optoisolator replaces transformer $T2$. The carrier method is not the most widespread in optically coupled isolation amplifiers because of the frequency response limitations of some IC optoisolators. Only recently have these problems been solved.

The more common direct approach is shown in Fig. 7-4. This circuit uses the same DC-to-DC converter to power the isolated stages as was used in other designs. It keeps $A1$ isolated from the AC power mains but is not used in the signal coupling process. In some designs, the high frequency carrier power supply is actually a separate block from the isolation amplifier.

The LED in the optoisolator is driven by the output of isolated amplifier $A1$. Transistor $Q1$ serves as a series switch to vary the light output of the LED proportional to the analog signal from $A1$. Transistor $Q1$ normally passes sufficient collector current to bias the LED into a linear portion of its operating curve. The output of the phototransistor is AC-coupled to the remaining amplifiers on the nonisolated side of the circuit, so that the offset condition created by the LED bias is eliminated.

Although not strictly speaking an isolated amplifier by the definition used herein, there is another category of optical isolation that is especially attractive for applications where the environment is too hostile for ordinary electronics. It is possible to use LED and phototransistor transmitter and

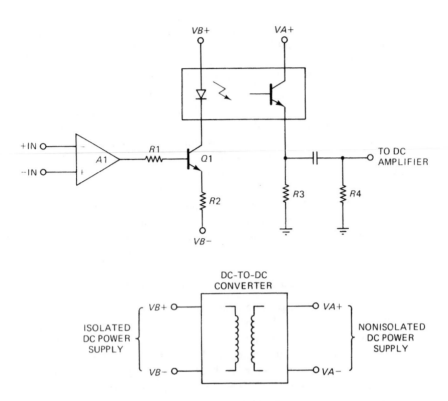

VB+   VA+

+IN
−IN

A1   R1   Q1

R2

VB−

R3   R4

TO DC
AMPLIFIER

DC-TO-DC
CONVERTER

ISOLATED
DC POWER
SUPPLY

VB+   VA+

VB−   VA−

NONISOLATED
DC POWER
SUPPLY

FIGURE 7-4   Optical coupling method for amplifier isolation.

receiver modules in a fiber optic system to provide isolation. A battery powered (or otherwise isolated) amplifier will sense the desired signal, convert to an AM or FM light signal, and transmit it down a length of fiber optic cable to a phototransistor receiver module. At that point the signal will be recovered and processed by the nonisolated electronics.

7-3.4 Current loading isolation amplifier methods

A unique 'current loading' isolation amplifier was used in the front end of an electrocardiograph (ECG) medical monitor that found widespread acceptance. A simplified schematic is shown in Fig. 7-5. Notice that there is no obvious coupling path for the signal between the isolated and nonisolated sides of the circuit.

The gain-of-24 isolated input preamplifier (A1) in Fig. 7-5 consists of a high input impedance operational amplifier. This amplifier is needed in order to interface with the very high source impedance normal to electrodes in ECG systems. The output of A1 is connected to the isolated −10 volt DC power supply through load resistor R1. This power supply is a DC-to-DC converter operating at 250 kHz. Transformer T1 provides isolation between the floating

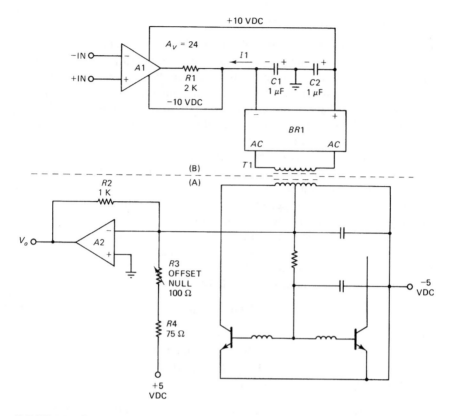

FIGURE 7-5 Current loading isolation method.

power supplies on the isolated B side of the circuit and the nonisolated A side of the circuit (which are AC-line powered).

An input signal causes the output of A1 to vary the current loading of the floating −10 volt DC power supply. Changing the current loading proportional to the analog input signal causes variation of the T1 primary current that is also proportional to the analog signal. This current variation is converted to a voltage variation by amplifier A2. An offset null control (R3) is provided in the A1 circuit to eliminate the offset at the output due to the quiescent current flowing when the analog input signal is zero. In that case, the current loading of T1 is constant, but still provides an offset to the A2 amplifier.

7-4 DESIGN EXAMPLES

In the sections below you will find several design examples that were selected to illustrate the applications of isolation amplifiers. These applications are drawn from medical, scientific and process engineering fields.

7-4.1 Cardiac output computer

The problems presented by most electronic signals acquisition situations are simple compared with the problems presented in measuring human cardiac output. The principal difference is that cardiac output is usually measured using an invasive surgical technique on living humans. This type of measurement is presented here in order to demonstrate a data acquisition technique that for the sake of safety almost absolutely requires an isolation amplifier to interface with the signal source.

Cardiac output (CO) is defined as the rate of blood volume pumped by the heart. The question being asked of the CO measurement is, 'how much blood is this person pumping per unit of time'. Cardiac output is measured in units of liters of blood per minute of time (l/min). In healthy adults CO typically reaches a value between 3 and 5 l/min.

A quantitative measure of cardiac output is the product of the stroke volume and the heart rate. The stroke volume is merely the volume of blood expelled from the heart ventricle (lower chamber) during a single contraction of the heart. Cardiac output is calculated from:

$$CO = V \times R \tag{7-1}$$

where:
    CO is the cardiac output in liters per minute (l/min)
    V is the stroke volume in liters per beat (l/beat)
    R is the heart rate in beats per minute (beat/min)

It is difficult, and usually impossible (except on animals in laboratory settings) to directly measure cardiac output using any technique that is directly based on the above equation. The main problem is obtaining good stroke volume data without excessive risk to the patient.

The *thermodilution method* of cardiac output measurement has become the standard indirect method for measuring cardiac output in clinical settings, and is also popular among laboratory scientists. Thermodilution technique forms the basis for most clinical and research cardiac output computers now on the market. One reason why thermodilution is preferred is that no poisonous injectates are used (as they are in radioactive or optical dye dilution methods), only ordinary medical intravenous (IV) solutions such as normal saline or 5% dextrose in water (D_5W) are used.

The thermodilution measurement of cardiac output is made using a special hollow catheter that is inserted into one of the patient's veins, usually on the right arm (the brachial vein is popular). The catheter is multi-lumened, and one of the lumens has its output hole several centimeters from the catheter tip. This proximal lumen is situated so that it is outside the heart (close to the input valve on the right atrium) when the tip is all the way through the heart, resting in the pulmonary artery (Fig. 7-6). Other lumens in the catheter output at the tip are used to measure pressures in the pulmonary artery in

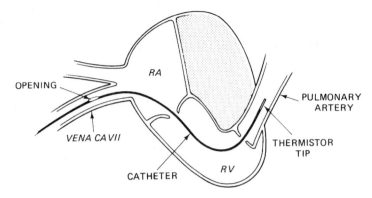

FIGURE 7-6 Measuring cardiac output requires a thermistor-tipped catheter to be inserted through the patient's veins, through the right side of the heart, into the pulmonary artery.

other procedures. A thermistor in the tip registers a resistance change with changes in blood temperature.

Most thermodilution cardiac output computers operate on a version of the following equation:

$$CO = \frac{64.8 \times C(t) \times V(i) \times [T(b) - T(i)]}{\int T(b)'dt} \qquad (7\text{-}2)$$

[NOTE: the mathematical integral symbol and the dt in the denominator of Eq. (7-2) indicates that intregration takes place, and that the result is the time-average of temperature $T(b)'$]

where:

CO is the cardiac output in liters per minute (l/min)

64.8 represents a collection of other constants and the conversion factor from seconds to minutes.

$C(t)$ is a constant that is supplied with the injectate catheter that accounts for the temperature rise in the portion of the outside of the victim's body.

$T(b)$ is the blood temperature in degrees Celcius.

$T(i)$ is the temperature of the injectate in degrees Celcius.

$T(b)'$ is the temperature of the blood as it changes due to mixing with the injectate.

The mathematical symbol in the denominator of Eq. (7-2) tells us that the temperature of the blood at the output side of the heart is integrated, i.e. the computer finds the time-average of the temperature as it changes.

EXAMPLE 7-1

A special cardiac output computer 'dummy catheter' test fixture enters a temperature signal that simulates a temperature change of 10°C for a period of 10 seconds. Find the expected reading during a test of the instrument if the following front panel settings are entered: $C(t) = 49.6$, $T(b) = 37°C$, injectate temperature $T(i)$ is $25°C$, and injectate volume $V(i)$ is 10 ml.

Solution

$$CO = \frac{64.9 \times C(t) \times V(i) \times [T(b) - T(i)]}{\int T(b)' \, dt} \text{ liters/min}$$

$$CO = \frac{(64.8)(49.6)\left[10 \text{ ml} \times \dfrac{1\,1}{1000 \text{ ml}}\right] \times [37°C - 25°C]}{(10°C)(10 \text{ s})} \text{ liters/min}$$

$$CO = \frac{3214 \times 0.01 \times 12}{100} \text{ liters/min} = 3.86 \text{ liters/min} \qquad \blacksquare$$

The thermistor in the end of the catheter is usually connected in a Wheatstone bridge circuit (see Fig. 7-7). The DC excitation of the bridge is critical. Either the short-term stability of this voltage must be very high, or a ratiometric method must be used to cancel excitation potential drift. In addition, it is also necessary to limit the bridge excitation potential to about 200 mV for reasons of safety to the patient (electrical leakage is especially dangerous because the thermistor in *inside the heart or pulmonary artery*). This low value of excitation voltage promotes both patient safety and thermistor stability through freedom from self-heating induced drift, even though imposing a greater burden on the amplifier design.

Figure 7-7 shows a simplified schematic of a typical cardiac output computer front-end circuit. The thermistor is in a Wheatstone bridge circuit consisting also of $R1$ through $R3$, with potentiometer $R5$ serving to balance

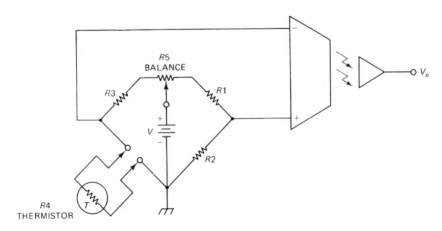

FIGURE 7-7 Wheatstone bridge forms the basis for thermistor measurement circuit at input of differential isolation amplifier.

the bridge. An autobalancing or zeroing method is sometimes used for this function. Those circuits use a digital-to-analog converter (DAC) to inject a current into one node of the bridge, and that current nulls the bridge circuit to zero. The physician waits a few minutes for the thermistor to equilibrate with blood temperature (usually it is in this condition by the time it is threaded through the venous system to the pulmonary artery). The output of the bridge, depending upon the design, is typically 1.2 to 2.5 mV/°C, with 1.8 mV/°C being quite common. This signal is amplified approximately 1000 times to 1 volt/°C by the preamplifier. This preamplifier is an isolated amplifier for reasons of patient safety. The output of this circuit, V_o, is used in the denominator of an equation as above.

The block diagram for a sample analog cardiac output computer is shown in Fig. 7-8. The front-end circuitry from Fig. 7-7 is in the blocks marked Bridge and Pre-amp. The isolator circuit is merely a buffer amplifier that permits V_o to be output to an analog paper chart recorder. Analysis of the waveshape reveals errors of technique, and thus explains odd readings that are not supported by other clinical facts. For this reason the physician demands an analog output. The temperature signal (V_o) is integrated, and then sent to an analog divider where it is combined with the temperature difference signal $[T(\text{b}) - T(\text{i})]$ and the constants (all represented by a single voltage). The low-pass filtered output of the analog multiplier is a measure of the cardiac output, and is displayed on a digital voltmeter.

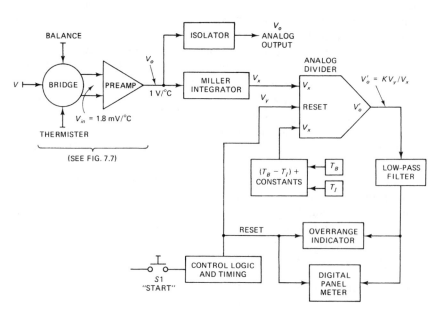

FIGURE 7-8   Block diagram of analog cardiac output computer.

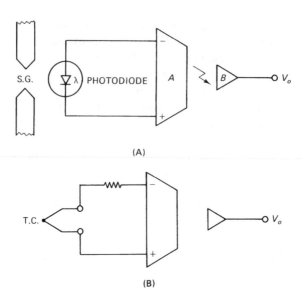

(A)

(B)

FIGURE 7-9　(A) Spark detection circuit; (B) thermocouple application.

7-4.2　Sensor isolation applications

Medical applications of isolation amplifiers are perhaps the best known because they contribute much to patient safety. Nonetheless, there is also a substantial body of non-medical applications for isolation amplifiers. These amplifiers are especially useful whenever the environment (e.g. high voltages) can adversely affect the device, or a conventional amplifier can perturb the signal source (certain transducers fall into this category). Figure 7-9 shows two situations where an isolation amplifier is interfaced to different forms of sensor. In Fig. 7-9A the sensor is a photodiode used to detect the existence of an electrical arc across a spark gap (SG). In some such systems strong electric fields can destroy conventional ground referenced systems.

Another sensor application is shown in Fig. 7-9B. A *thermocouple* (TC) is a temperature transducer consisting of two dissimilar metals forming a junction. If the metals have different work functions, then an electric potential is generated across the ends of the wires that is proportional to both the work function difference and the junction temperature. Ordinarily, the TC can be connected to any amplifier. But in certain cases the local electrical environment is too adverse for the electronic components. An example is a high voltage melting vessel for smelting metals. It is necessary to know and control the temperature, but the vessel floats electrically at a high AC potential.

7-4.3　Electrophoresis column

Electrophoresis is a process used in biochemistry laboratories to separate proteins from biological fluids. In an electrophoresis column (Fig. 7-10), a

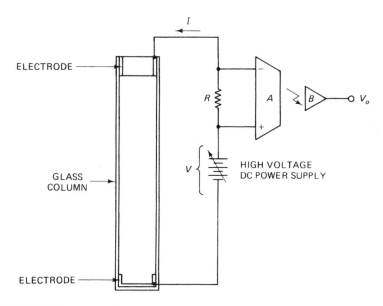

FIGURE 7-10 Electrophoresis column application for isolation amplifier.

pair of platinum electrodes connected to a constant-current, high voltage DC power supply creates an electrical field between the top and bottom of the column. Various protein molecules within the fluid react very specifically to the field and will migrate to, and settle at, a specific level in the field (hence height within the column). The chemist can then extract the particular protein of interest if the height of the column and the potential gradient are known.

It is necessary to continuously monitor the current flowing in the column. If computer data logging is used, then an isolation amplifier may be required to keep the high voltage of the circuit from the computer. A series resistor is inserted into the column circuit, and the voltage drop across it is proportional to the current flowing (Ohm's law). The differential inputs of the isolated side of the amplifier are bridged across this resistor to indirectly measure the current drain.

7-5 A COMMERCIAL PRODUCT EXAMPLE

Figure 7-11 shows the circuit of an isolation amplifier based on the Burr-Brown 3652 device.

The DC power for both the isolated and nonisolated sections of the 3652 is provided by the 722 dual DC-to-DC converter. This device produces two independent ±15 VDC supplies that are each isolated from the 50/60 Hz AC power mains and from each other. The 722 device is powered from a +12 VDC source that is derived from the AC power mains. In some cases, the nonisolated section (which is connected to the output terminal) is powered

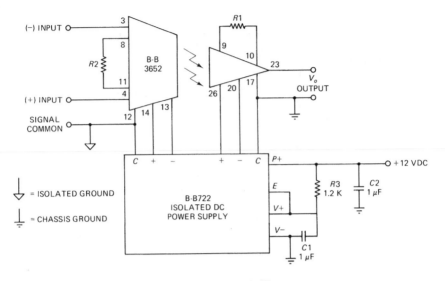

FIGURE 7-11 Typical circuit for the Burr-Brown 3652 isolation amplifier.

from a bipolar DC power supply that is derived from the 50/60 Hz AC mains, such as a ±12 VDC or ±15 VDC supply. In no instance, however, should the isolated DC power supplies be derived from the AC power mains.

There are two separate ground systems in this circuit, symbolized by the small triangle and the regular three-bar 'chassis' ground symbol. The isolated ground is not connected to either the DC power supply ground/common, or the chassis ground. It is kept floating at all times, and becomes the signal common for the input signal source.

The gain of the circuit is approximately:

$$GAIN = \frac{1\,000\,000}{R1 + R2 + 115} \tag{7-3}$$

In most design cases, the issue is the unknown values of the gain setting resistors. We can rearrange the equation above to solve for $(R1 + R2)$:

$$(R1 + R2) = \frac{1\,000\,000 - (115 - GAIN)}{GAIN} \tag{7-4}$$

where:
  $R1$ and $R2$ are in ohms (Ω)
  $GAIN$ is the voltage gain desired

EXAMPLE 7-2

An amplifier requires a differential voltage gain of 1000. What combination of $R1$ and $R2$ will provide that gain figure?

Solution

If $GAIN = 1000$

$$R1 + R2 = \frac{1\,000\,000 - (115 \times 1000)}{1000}$$

$$R1 + R2 = \frac{1\,000\,000 - 115\,000}{1000} = \frac{885\,000}{1000} = 885 \text{ ohms}$$

In this case, we need some combination of $R1$ and $R2$ that adds to 885 ohms. The value 440 ohms is 'standard', and will result in only a tiny gain error if used. ∎

7-6 CONCLUSION

Although the isolation amplifier is considerably more expensive than common IC linear amplifiers, there are applications where these amplifiers are absolutely critical. Wherever the instrument could cause injury to a human, or wherever the environment is such that the electronics must be isolated as far as possible, the isolation amplifier is the device of choice (at least in the front-end).

7-7 SUMMARY

1. The purpose of the isolation amplifiers is to increase the resistance between the inputs and either the output or the AC power lines to as high a number as possible. Isolation impedances on the order of 10^{12} ohms are possible.

2. There are four basic approaches to the design of isolation amplifiers: *battery powered, carrier operated, optically coupled* and *current loading.*

3. Typical examples of isolation amplifier applications include: medical biopotentials amplifiers (ECG, EEG, and EMG), medical cardiac output computers, intra-aortic pressure meters, sensor or transducer isolation, and isolation of process control electronics from harsh or hostile electrical environments.

7-8 RECAPITULATION

Now return to the objectives and Pre-quiz questions at the beginning of the chapter and see how well you can answer them. If you cannot answer certain questions, place a check mark to each and review the appropriate parts of the text. Next, try to answer the questions and work the problems below, using the same procedure.

7-9 STUDENT EXERCISES

[NOTE: In the exercises below limit the -3 dB frequency response of the analog amplifier to a range of 1 to 100 Hz.]

1. Design, build and test an isolation amplifier based on optical coupling. Assume that the analog input signal will directly modulate the phototransmitter.
2. Design an isolation amplifier based on fiber optic coupling. State in your laboratory report how this design might be superior to others in certain applications.
3. Design, build and test an isolation amplifier based on optical coupling. In this exercise use a voltage controlled oscillator at some convenient frequency to frequency modulate the light beam.
4. Design, build and test an isolation amplifier of the carrier-operated type. In order to use ordinary audio components limit the carrier frequency to 20 kHz.

7-10 QUESTIONS AND PROBLEMS

1. Define isolation amplifier.
2. List several applications for isolation amplifiers, and explain why each requires the isolation factor.
3. What is the goal of isolation amplifier design.
4. List the connections typically found on an isolation amplifier that uses differential inputs.
5. List four methods or approaches for designing an isolation amplifier.
6. An amplifier has isolated input and output sections, but uses the same common or ground connection for both sections. Is this amplifier isolated when powered from 110 volt AC power lines?
7. A synchronous detector is used on a _____ -operated isolation amplifier.
8. A Burr-Brown 3652 is used in an application requiring an isolation amplifier. What is the gain if both gain setting resistors are 600 ohms?
9. Use a Burr-Brown 3652 to design an isolation amplifier front-end for a cardiac output computer. Select resistor values that will yield a gain of 400 for the input amplifier.
10. In a carrier type isolation amplifier, a voltage controlled oscillator is _____ modulated by the input signal before being applied to an optocoupler.
11. Why are isolation amplifiers used in the front-ends of ECG preamplifiers and cardiac output computers.
12. A special cardiac output computer test fixture simulates the thermistor catheter resistance change representing a $20°C$ temperature change. Calculate the indicated cardiac output reading that would be expected if the temperature change is inserted in the circuit for 7 seconds, and the following constants are entered into the computer: $C(t) = 64.5$, $T(b) = 37°C$, $T(i) = 20°C$, and $V(i) = 20$ ml.
13. Why are isolation amplifiers sometimes used in nonmedical instrumentation applications?

CHAPTER 8

Other IC linear amplifiers

OBJECTIVES

1. Learn basic theory for the operational transconductance amplifier (OTA).
2. Learn typical applications for the OTA.
3. Learn the basic theory for the current difference (a.k.a. 'Norton') amplifier (CDA).
4. Learn typical applications for the CDA.

8-1   PRE-QUIZ

These questions test your prior knowledge of the material in this chapter. Try answering them before you read the chapter. Look for the answers (especially those you answered incorrectly) as you read the text. After you have finished studying the chapter try answering these questions again, and those at the end of the chapter (see Section 8-8).

1. Write the transfer equation for the operational transconductance amplifier.
2. Write the transfer equation for the current difference amplifier.
3. How can an OTA be used as a voltage amplifier?
4. Design a low output impedance OTA voltage amplifier circuit.

8-2   INTRODUCTION

The operational amplifier is a simple voltage amplifier with the simple transfer function $A_v = V_o/V_{in}$. While that type of device is the most commonly used form of IC linear amplifier, there are cases where another form of linear amplifier is needed. In this chapter we will take a look at two other popular forms of IC linear amplifiers: the *operational transconductance amplifier* (OTA) and the *current difference amplifier* (CDA), also sometimes called the *Norton amplifier*. These devices are not likely to replace the operational amplifier, but nonetheless find their own niche in the integrated electronics marketplace.

8-3 OPERATIONAL TRANSCONDUCTANCE IC AMPLIFIERS (OTA)

In this section we will take a look at a linear IC device called the operational transconductance amplifier (OTA). The OTA is based on a transfer function that *relates an output current to an input voltage*. In other words:

$$G_m = \frac{I_o}{V_{in}} \tag{8-1}$$

where:
G_m is the transconductance in siemens or microsiemens
I_o is the output current
V_{in} is the input voltage

The operational transconductance amplifier equivalent circuit is shown in Fig. 8-1. The differential input circuit is similar to the input circuit of

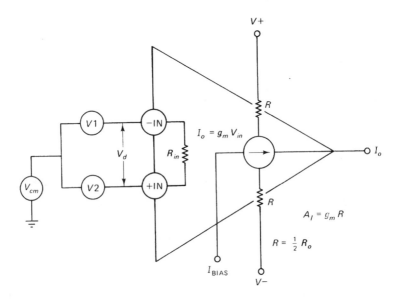

FIGURE 8-1   Equivalent circuit for operational transconductance amplifier (OTA).

the operational amplifier because each are differential voltage inputs (−IN and +IN). The input voltages are differential signal V_d (i.e. $V2 - V1$) and common mode signal V_{cm}. The output side of the amplifier, however, is a *current source* that produces an output current, I_o, which is proportional to the gain and the input voltage. The current gain (A_{gm}) of this circuit is a function of the transconductance (I_o/V_{in}) and the load resistance, R:

$$A_{gm} = G_m \times R \qquad (8\text{-}2)$$

where:

A_{gm} is the gain
G_m is the transconductance (I_o/V_{in})
R is the load resistance (one-half the output resistance R_o).

[NOTE: G_m and R must be expressed in equivalent reciprocal units. In other words, when G_m is in siemens then R must be in ohms. Likewise, millisiemens and milliohms; microsiemens microohms are also paired.]

Perhaps the most common commercial version of the operational transconductance amplifier (OTA) is the CA-3080, CA-3080A and CA-3060 devices. The CA-3080 devices are available in the eight-pin metal IC package using the pinouts shown in Fig. 8-2. The CA-3080 devices will operate over DC power supply voltages from ±2 volts to ±15 volts, with adjustable power consumption of 10 microwatts to 30 milliwatts. The gain is 0 to the product G_mR. The input voltage spread is ±5 volts. The bias current can be set as high as 2 mA.

Note that the pinouts for the CA-3080 device are industry standard operational amplifier pinouts, except for the bias current applied to pin number 5:

$V-$ on pin 4
$V+$ on pin 7

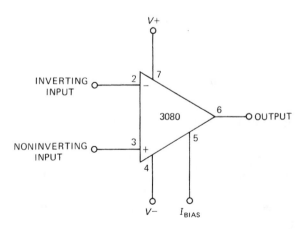

FIGURE 8-2   Schematic symbol and pinouts for the CA-3080 OTA.

Inverting input (−IN) on pin 2
Noninverting input (+IN) on pin 3
Output on pin 6

The operating parameters of the operational transconductance amplifier are set by the bias current (I_{bias}). For example, on the CA-3080 device the transconductance is 19.2 times higher than the bias current:

$$G_m = 19.2 \times I_{bias} \tag{8-3}$$

where:
  G_m is in millisiemens
  I_{bias} is in milliamperes

In many actual design cases you will know the required value of G_m from knowledge of I_o/V_{in}, so the G_m required can be set by adjusting the bias current. In those cases, the I_{bias} is found by rewriting the above expression:

$$I_{bias} = \frac{G_m}{19.2} \tag{8-4}$$

The CA-3080 output resistance is also a function of the bias current:

$$R_o = \frac{7.5}{I_{bias}} \tag{8-5}$$

where:
  R_o is the output resistance in megohms (MΩ)
  I_{bias} is the bias current in milliamperes (mA)

EXAMPLE 8-1

What is the output resistance (R_o) when the bias current is 500 μA (i.e. 0.5 mA)?

Solution

$$R_o = \frac{7.5}{I_{bias}}$$

$$R_o = \frac{7.5}{0.5 \text{ mA}} = 15 \text{ MΩ} \qquad \blacksquare$$

8-3.1 Voltage amplifier from the OTA

The OTA is a current-output device, but can be used as a voltage amplifier when one of the circuit strategies of Fig. 8-3 are used. The simplest method is the resistor load shown in Fig. 8-3A. Because the output of the OTA is a current (I_o), we can pass this current through a resistor ($R1$) to create a voltage drop. The value of the voltage drop (and the output voltage, V_o) is found from Ohm's law:

$$V_o = I_o \times R1 \tag{8-6}$$

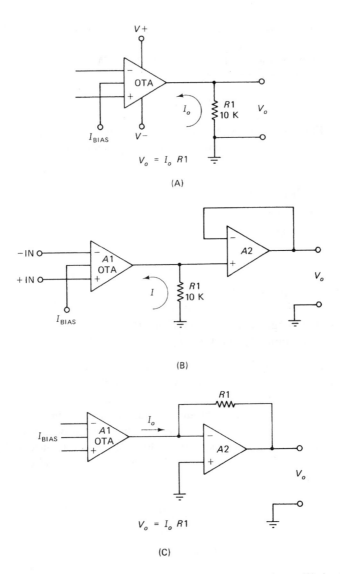

FIGURE 8-3 Circuits for converting output current to output voltage: (A) simple resistor load; (B) resistor load buffered by op-amp; (C) direct amplification using op-amp.

A problem with this circuit is that the source impedance is very high, being equal to the value of $R1$. In the example shown in Fig. 8-3A, the output impedance is 10 kohms. We can overcome this problem by adding a unity gain noninverting operational amplifier such as $A2$ in Fig. 8-3B. The output voltage in this case is the same as for the nonamplifier version: $I_0 \times R1$, although the output impedance is very low.

Another form of low-impedance output circuit is shown in Fig. 8-3C. This form uses the inverting follower configuration of the operational amplifier ($A2$). The output voltage is the product of the OTA output current (I_o) and the op-amp feedback resistor ($R1$):

$$V_o = I_o \times R1 \qquad (8\text{-}7)$$

In both Figs 8-3B and 8-3C the output impedance is equal to the operational amplifier output impedance, which is typically something less than 100 ohms.

8-3.2 OTA applications

An 'analog multiplier' is a circuit that produces a voltage that is the product of two input voltages:

$$V_o = rV_xV_y \qquad (8\text{-}8)$$

where:

V_o is the output voltage
V_x is the voltage applied to the X-input
V_y is the voltage applied to the Y-input
r is a proportionality constant

There are many applications for the multiplier circuit, even though some of them are now generally performed in a digital computer or processor. Immediately one can see instrumentation applications, even in this era of computers. Also possible are amplitude modulation and demodulation tasks for analog multipliers. Operational transconductance amplifiers can be used to make two-quadrant and four-quadrant analog multipliers. Consider Fig. 8-4, which is an analog XY multiplier based on the CA-3060 quad OTA. Recall from Eq. (8-1) that:

$$G_m = \frac{I_o}{V_{in}}$$

Therefore:

$$I_{o1} = -V_xG_{M1} \qquad (8\text{-}9)$$

and

$$I_{o2} = +V_xG_{M2} \qquad (8\text{-}10)$$

Because the value of R_o for each OTA is very large compared with the load, we can simply sum the two output currents:

$$I_o = I_{o2} + I_{o1} \qquad (8\text{-}11)$$

By $V_o = I_oR_L$:

$$V_o = (I_{o2} + I_{o1})R_L \qquad (8\text{-}12)$$

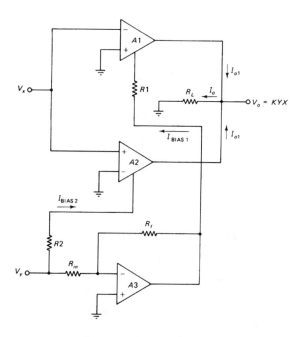

FIGURE 8-4   OTA based analog multiplier circuit.

$$V_o = [(+V_x)(G_{M2}) + (-V_xG_{M1})]R_L \qquad (8-13)$$

$$V_o = (G_{M2} - G_{M1})V_xR_L \qquad (8-14)$$

Recall that $G_m = kI_{bias}$. For amplifier $A2$ we know that:

$$I_{bias} = \frac{(V-) + (V+)}{R1} \qquad (8-15)$$

Therefore,

$$G_{M2} = K[(V-) + (V_x)] \qquad (8-16)$$

Using a similar line of reasoning we arrive at:

$$G_{M1} = K[(V-) + (V_y)] \qquad (8-17)$$

Combining Eqs (8-14), (8-16) and (8-17) yields:

$$V_o = V_xKR_L[(V-) + (V_x)] - [(V-) - (V_y)] \qquad (8-18)$$

or, after simplifying terms:

$$V_o = 2KR_LV_xV_y \qquad (8-19)$$

Equation (8-19) is the transfer equation for Fig. 8-4. It has the same form as Eq. (8-8) in which $r = 2KR_L$.

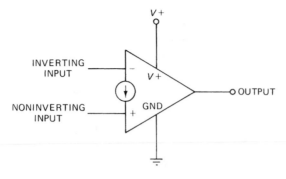

FIGURE 8-5 Schematic symbol for current difference amplifier (CDA).

8-4 CURRENT DIFFERENCE AMPLIFIERS _____

The current difference amplifier (CDA), also called the Norton amplifier, is another non-operational linear IC amplifier that performs similarly to the op-amp, but not exactly the same. The CDA has certain features that make it uniquely useful for certain applications. One place where the CDA is more useful than the operational amplifier is in circuits that process AC signals, but are limited to a single-polarity DC power supply. For example, in automotive electronics equipment limited to a single 12 to 14.4 volt DC battery power supply that uses the car chassis for negative common return. There are other cases where the linear IC amplifier is but a minor feature of the circuit, most of which operates from a single DC power supply. It might be wasteful in such circuits to use the operational amplifier. We would either have to bias the operational amplifier with an external resistor network, or provide a second DC power supply.

The normal circuit symbol for the CDA is shown in Fig. 8-5. This symbol looks much like the regular op-amp symbol, except that a current source symbol is placed along the side opposite the apex. This symbol is typically used for several products such as the LM-3900 device, which is a quad Norton amplifier. You may sometimes find schematics where the op-amp symbol is used for the CDA, but that is technically incorrect usage.

8-4.1 CDA circuit configuration

The input circuit of the CDA differs radically from the operational amplifier. Recall that the op-amp used a differential input amplifier driven from a constant current source supplying the collector–emitter current. The CDA is quite different, however, as can be seen in Fig. 8-6.

The overall circuit of a typical CDA is shown in Fig. 8-6A, while an alternative form of the input circuit is shown in Fig. 8-6B. Transistor $Q7$ in Fig. 8-6A forms the output transistor, while $Q5$ is the driver. Both the NPN output transistor and the PNP driver transistor operate in the emitter

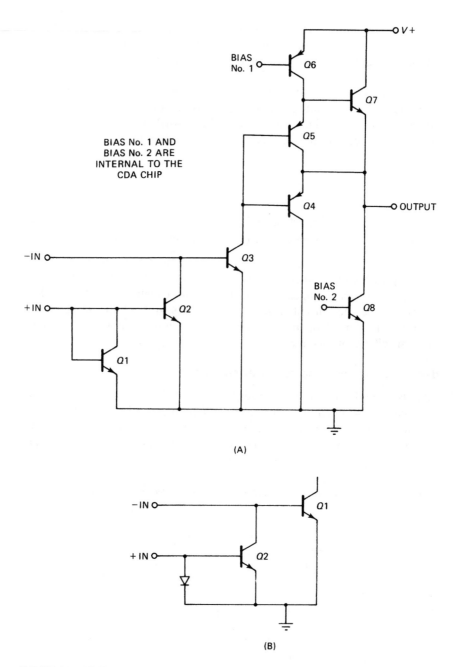

FIGURE 8-6   (A) Typical circuit for CDA; (B) input circuit of CDA showing current mirror (Q2).

follower configuration. Transistors $Q4$, $Q5$, $Q6$ and $Q8$ are connected to serve as current sources. The input transistor is $Q3$, and it operates in the common emitter configuration. The base of $Q3$ forms the inverting ($-$IN) input for the CDA.

The noninverting input of the CDA is formed with a 'current mirror' transistor, $Q1$ (transistor $Q1$ in Fig. 8-6A is 'diode connected' and serves exactly the same function as diode $D1$ in Fig. 8-6B). The dynamic resistance offered by the current mirror transistor ($Q2$) is given by:

$$r = \frac{26}{I_b} \tag{8-20}$$

where:
  r is the dynamic resistance of $Q2$ in ohms (Ω)
  I_b is the base bias current of $Q3$ in milliamperes (mA)

Equation (8-20) is used only at normal room temperature because I_b will vary with wide temperature excursions. For most common applications, however, the room temperature version of the equation will suffice. Data sheets for specific current difference amplifiers give additional details for amplifiers that must operate outside of the relatively narrow temperature range specified for the simplified equation.

8-4.2   CDA inverting follower circuits

Like its operational amplifier cousins, the CDA can be configured in either inverting or noninverting follower configurations. The inverting follower is shown in Fig. 8-7. In many respects this circuit is very similar to operational amplifiers. The voltage gain of the circuit is set approximately by the ratio

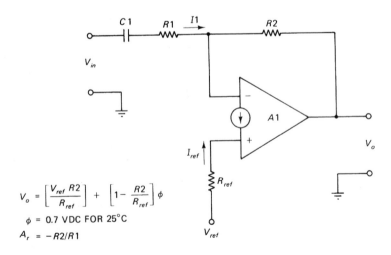

$$V_o = \left[\frac{V_{ref} \, R2}{R_{ref}} \right] + \left[1 - \frac{R2}{R_{ref}} \right] \phi$$

$\phi = 0.7$ VDC FOR 25°C

$A_r = -R2/R1$

FIGURE 8-7   CDA inverting follower circuit.

of the feedback to the input resistor:

$$A_v = \frac{-R2}{R1} \qquad (8\text{-}21)$$

where:

A_v is the voltage gain
R_2 is the feedback resistance
$R1$ is the input resistance

(Note: $R1$ and $R2$ are in the same units. The minus sign indicates that a 180° phase reversal occurs between input and output signals.)

We must provide a bias to the current mirror transistor (i.e. $Q2$ in Figure 8-6A), so resistor R_{ref} is connected in series with the noninverting input of the CDA and a reference voltage source, V_{ref}. In many practical circuits the reference voltage source is merely the $V+$ supply used for the CDA. In other cases, however, some other potential might be required, or alternatively the reference current is required to be regulated more tightly (or with less noise) than the supply voltage. Ordinarily we set the reference current to some convenient value between 5 μA and 100 μA. For $V+$ power supply values of $+12$ VDC, for example, it is common to find a 1 megohm resistor used for R_{ref}. In that case, the input reference current is: $I_{ref} = 12\ \text{V}/1\,000\,000\ \Omega = 12\ \mu\text{A}$.

A constraint placed on CDAs is that the input resistor ($R1$) used to set gain must be high compared with the value of the current mirror dynamic resistance (Eq. (8-20) above). The CDA becomes nonlinear (i.e. distorts the input signal) if the input resistor value approaches the current mirror resistance, r. In that case, the voltage gain is not $-R2/R1$, but rather:

$$A_v = \frac{R2}{R1 + r} \qquad (8\text{-}22)$$

where:

A_v is the voltage gain
$R2$ is the feedback resistance
$R1$ is the input resistance
r is the current mirror resistance

Equation (8-22) essentially reduces to Eq. (8-21) when we can force $R1$ to be very much larger than r. This goal is easily achieved in most circuits because r is tiny (work a few examples with normal bias currents in Eq. (8-22)).

The output voltage of the CDA will exhibit an offset potential even when the AC input signal is zero. This potential is given by:

$$V_o = \left[\frac{V_{ref}R2}{R_{ref}}\right] + \left[1 - \frac{R2}{R_{ref}}\right]\Psi \qquad (8\text{-}23)$$

where:

V_o is the output potential in volts (V)

V_{ref} is the reference potential in volts, usually $V+$ (V)

$R2$ is the feedback resistance in ohms (Ω)

R_{ref} is the current mirror bias resistance in ohms (Ω)

Ψ is a temperature-dependent factor (0.70 volts for room temperature)

The capacitor in series with the input circuitry has the effect of limiting the low end frequency response. The -3 dB cutoff frequency is a function of the value of this capacitor and the input resistance, $R1$. This frequency, F, is given by:

$$F_{Hz} = \frac{1\,000\,000}{2\pi R1 C1} \qquad (8\text{-}24)$$

where:

F is the lower end -3 dB frequency in hertz (Hz)

$R1$ is in ohms (Ω)

$C1$ is in microfarads (μF)

In some CDA circuits there is also a capacitor in series with the output terminal. The purpose of that capacitor is to prevent the DC offset that is inherent in this type of circuit from affecting following circuits. The output capacitor will also limit the low end frequency response. The same form of equation (i.e. Eq. (8-25)) is used to determine this frequency, but using the input resistance of the load as the R term.

As is often the case with equations presented in electronics books, Eq. (8-24) is not necessarily in the most useful form. In most cases, you will know the input resistance ($R1$) from the application. It is typically not less than ten times the source impedance, and forms part of the gain equation. Consideration of driving source impedance and voltage gain tends to determine the value of $R1$. The required low end frequency response is usually determined from the application. You generally know (or can find out) that frequency spectrum of input signals. From the lower limit of the frequency spectrum you can determine the value of F. Thus, you will determine F and $R1$ from other considerations than the circuit. You therefore need a version of Eq. (8-24) that will allow us to calculate capacitor $C1$:

$$C1 = \frac{1\,000\,000}{2\pi R1 F} \qquad (8\text{-}25)$$

(all terms as defined for Eq. (8-24)).

DESIGN EXAMPLE

Design a gain-of-100 AC amplifier based on a CDA in which the input impedance is at least 10 kohms and the -3 dB frequency response is 3 Hz or lower. Assume DC power supplies of ±15 volts DC (see Fig. 8-8 for final circuit).

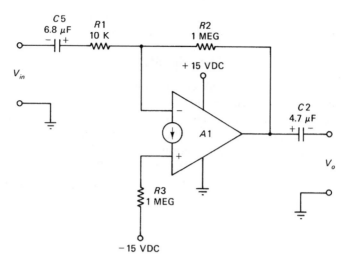

FIGURE 8-8 Practical CDA amplifier.

Solution

1. Set the reference current to the noninverting input to a value between 5 and 100 μA. Select 15 μA.

$$R3 = V/I_{ref}$$

$$R3 = (15 \text{ VDC})/(0.000015 \text{ A}) = 1\,000\,000 \text{ ohms}$$

2. Set the gain resistors. $R1$ can be $10\,000$ ohms in order to meet the input impedance requirement. From $A_v = R2/R1$, we know that $R2 = A_v \times R1$.

$$A_v = A_v \times R1$$

$$A_v = (100) \times (10\,000 \ \Omega) = 1\,000\,000 \ \Omega$$

3. Find the value of input capacitor $C1$ when $F_{3 \text{ dB}}$ is 3 Hz.

$$C1 = \frac{1\,000\,000}{2\pi R1 F}$$

$$C = \frac{1\,000\,000}{(2)(3.14)(10^4 \ \Omega)(3 \text{ Hz})}$$

$$C1 = \frac{1\,000\,000}{188\,400} = 5.3 \ \mu F$$

Because 5.3 μF is a nonstandard value, we select the next higher standard value (which is 6.8 μF).

The value of output capacitor $C2$ can be arbitrarily set to 6.8 μF if the load impedance is $10\,000$ ohms. If the load impedance is higher than that value, then use 4.7 μF or 6.8 μF. If the load impedance is very much higher than $10\,000$ ohms, or is lower than 10 000 ohms, then calculate the value using the same equation as for $C1$ but with the load resistance substituted for the input impedance. ∎

8-4.3 Noninverting amplifier circuits

The noninverting amplifier CDA configuration is shown in Fig. 8-9. This circuit retains the reference current bias applied to the noninverting input, but rearranges some of the other components. As in the case of the inverting amplifier configuration, the noninverting amplifier uses $R2$ to provide negative feedback between the output terminal and the inverting input. Unlike the inverting CDA circuit, however, input resistor $R1$ is connected in series with the noninverting input. The gain of the noninverting CDA amplifier is given by:

$$A_v = \frac{R2}{\left[\dfrac{26R1}{I_{\text{ref}}}\right]} \tag{8-26}$$

where:

    A_v is the voltage gain

    $R1$ and $R2$ are in ohms (Ω)

    I_{ref} is the bias current in milliamperes (mA)

The reference current, I_{ref}, is set to a value between 5 µA and 100 µA (i.e. 0.005 to 0.1 mA). Unlike the situation in the inverting amplifier, the value of this current is partially responsible for setting the gain of the circuit. Some clever designers have even used this current as a limited gain control for some CDA stages. The value of the resistor that provides the reference current (R_{ref}) is set by Ohms law, considering the required value of reference

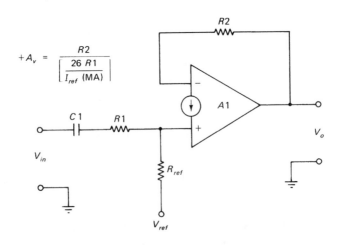

FIGURE 8-9 Noninverting follower CDA circuit.

current and the reference voltage, V_{ref}. In most common applications, the reference voltage is merely one of the supply voltages. The value of R_{ref} is determined from:

$$R_{ref} = \frac{V_{ref}}{I_{ref}} \qquad (8\text{-}27)$$

where:

R_{ref} is the reference resistor in ohms (Ω)

V_{ref} is the reference potential in volts (V)

I_{ref} is the reference current in amperes (note: 1 μA $= 0.000001$ A)

The value of input impedance is approximately equal to $R1$, provided that $R1$ is much higher than the dynamic resistance of the current mirror inside the CDA (which is typically the case).

As was true in the inverting follower case, the input capacitor ($C1$) sets the low end frequency response of the amplifier. The -3 dB frequency is given by exactly the same equation as for the inverting case (see Eqs (8-25) and (8-26)).

Figure 8-10 shows a modification of the noninverting follower circuit that allows for a noisy reference source. This type of amplifier circuit might be used where the DC power supplies that are used for the reference voltage are electrically noisy. Such noise could come from other stages in the circuit, or from outside sources.

The purpose in Fig. 8-10 is to form a reference voltage from a resistor voltage divider circuit consisting of $R3$ and $R4$. The value of V_{ref} will be:

$$V_{ref} = \frac{(V+)R3}{R3 + R4} \qquad (8\text{-}28)$$

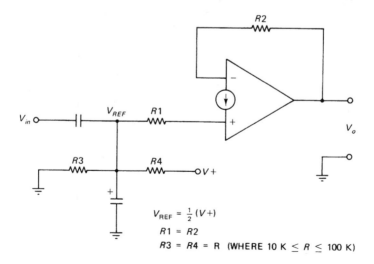

FIGURE 8-10   Noninverting follower circuit that limits reference source noise.

where:

V_{ref} is the reference potential in volts (V)

$V+$ is the supply potential in volts (V)

$R3$ is in ohms (Ω)

Inspection of Eq. (8-28) reveals that $V_{ref} = (V+)/2$ when $R3 = R4$, which is the usual case in practical circuits. The reference current is:

$$I_{ref} = \frac{V_{ref}}{R1} \qquad (8\text{-}29)$$

8-4.4 Super-gain amplifier

There is a practical limit to voltage gain using standard resistor values and standard circuit configurations (a similar problem also exists for operational amplifiers). In Fig. 8-11 we see a means for overcoming the limitations. This supergain amplifier circuit forms a noninverting follower in a manner similar to the earlier circuit, except that feedback resistor $R2$ is driven from an output voltage divider network rather than directly from the output terminal of the

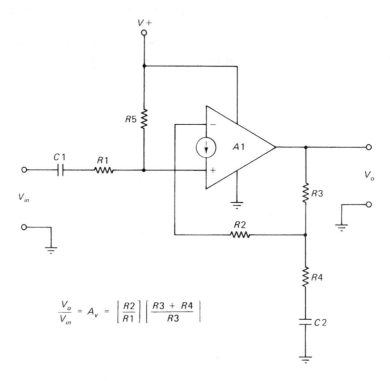

FIGURE 8-11   CDA super-gain amplifier circuit.

CDA. The voltage gain of the circuit of Fig. 8-11 is given by:

$$A_v = \left[\frac{R2}{R1}\right]\left[\frac{R3+R4}{R3}\right] \qquad (8\text{-}30)$$

EXAMPLE 8-2

Find the gain of the circuit shown in Fig. 8-6 if $R1 = 10$ kohms, $R2 = 100$ kohms, $R3 = 10$ kohms and $R4 = 100$ kohms.

Solution

$$A_v = \left[\frac{R2}{R1}\right]\left[\frac{R3+R4}{R3}\right]$$

$$A_v = \left[\frac{100\ \text{k}\Omega}{10\ \text{k}\Omega}\right]\left[\frac{100\ \text{k}\Omega + 10\ \text{k}\Omega}{10\ \text{k}\Omega}\right]$$

$$A_v = (10) \times (110/10) = 10 \times 11 = 110 \qquad \blacksquare$$

Capacitor $C1$ is set using the same equation that was used previously, while $C2$ is set to have a capacitive reactance of $R4/10$ at the lowest frequency of operation (in other words, the low-end -3 dB point).

8-4.5 CDA differential amplifiers

A differential amplifier is one that will produce an output that is proportional to the gain and the difference between potentials applied to the inverting and noninverting inputs. Figure 8-12 shows the circuit of a CDA differential amplifier. It is similar to the operational amplifier version of this simple circuit in several respects. For example, the two input resistors are equal and the differential voltage gain is the ratio of the negative feedback resistor ($R3$) and the input resistor:

$$A_v = \frac{R3}{R2} \qquad (8\text{-}31)$$

if $R1 = R2$ and $R3 = R4 = R5$.

The input impedance (differential) of this circuit is twice the value of the input resistances: $R_{in} = R1 + R2$. The bias current is provided through the two series resistors, $R4$ and $R5$.

Assuming that $R1 = R2 = R$, and $C1 = C2 = C$, we can calculate the low-end -3 dB frequency from the equation:

$$F = \frac{1\,000\,000}{2\pi RC} \qquad (8\text{-}32)$$

where:
F is the -3 dB frequency in hertz (Hz)
R is in ohms (Ω)
C is in microfarads (μF)

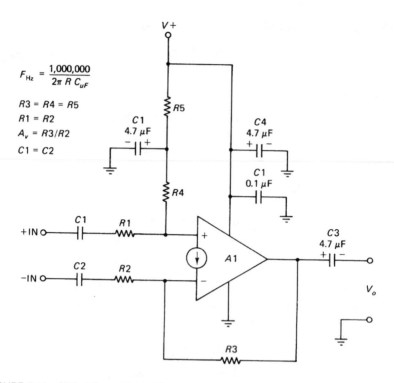

$$F_{Hz} = \frac{1,000,000}{2\pi \, R \, C_{uF}}$$

R3 = R4 = R5
R1 = R2
A_v = R3/R2
C1 = C2

FIGURE 8-12 CDA differential amplifier.

When this circuit is used as a 600-ohm line receiver, we can make the two input resistors 330 ohms each (or 270 ohms) if a small mismatch can be tolerated. Ideally, the input resistors will be 300 ohms each (which may require two resistors for each input resistor).

8-4.6 AC mixer/summer circuits

The CDA mixer or summer circuit is shown in Figure 8-13. This circuit is used to combine two or more inputs into one channel. The basic circuit is an inverting follower. Each input sees a gain that is the quotient of the feedback resistor to its input resistor:

$$A_{v1} = \frac{-R2}{R3} \qquad (8\text{-}33)$$

$$A_{v2} = \frac{-R2}{R4} \qquad (8\text{-}34)$$

$$A_{v3} = \frac{-R2}{R5} \qquad (8\text{-}35)$$

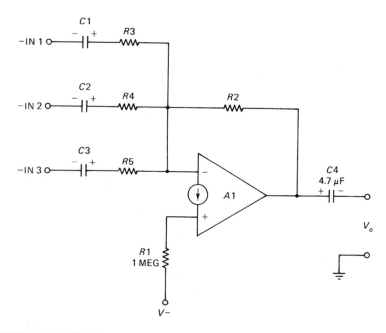

FIGURE 8-13   Multiple input inverting follower amplifier.

From Eqs (8-33) through (8-35) we can deduce that the output voltage is found from:

$$V_o = R2 \left[\frac{V1}{R3} + \frac{V2}{R4} + \frac{V3}{R5} \right] \qquad (8\text{-}36)$$

The frequency response of each channel is found from the usual equation for −3 dB frequency:

$$F = \frac{1\,000\,000}{2\pi RC}$$

where:

F is the −3 dB frequency in hertz (Hz)
R is the input resistance ($R3$, $R4$ or $R5$) in ohms (Ω)
C is the input capacitance ($C1$, $C2$ or $C3$) in microfarads (μF)

8-4.7   Differential output 600-ohm line driver amplifier: a circuit example

The 600-ohm line used in broadcasting electronics and professional audio recording requires either a center-tapped output transformer, or a linear amplifier with a push-pull output to drive the line. Figure 8-14 shows the circuit of a CDA 600-ohm line driver amplifier. This circuit basically consists of two

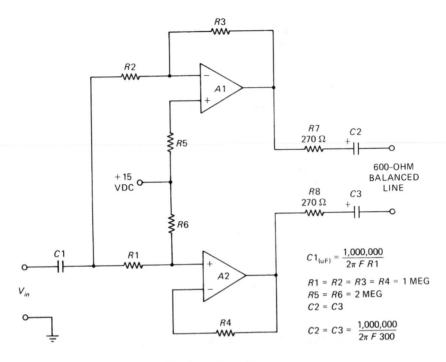

FIGURE 8-14   Driver circuit for 600-ohm balanced line.

separate amplifiers, one an inverting follower and the other a noninverting follower.

The bias resistors ($R5$ and $R6$) are set to provide a small bias current in the 5 to 100 µA. This current is found from Ohm's law, $I_{ref} = (V+)/R$. In the example shown in Fig. 8-14 the resistors are set to 2 megohms for a supply voltage of +15 volts DC.

The capacitors in the circuit set the low-end −3 dB point in the frequency response curve. These capacitor values are set from the following:

Assuming that $R1 = R2 = R3 = R4$, and $R5 = R6$:

$$C1 = \frac{1\,000\,000}{2\pi F R1} \tag{8-37}$$

Assuming $C2 = C3$:

$$C2 = \frac{1\,000\,000}{2\pi F 300} \tag{8-38}$$

where:
   $C1$ and $C2$ are in microfarads (µF)
   F is in hertz (Hz)
   $R1$ is in ohms (Ω)

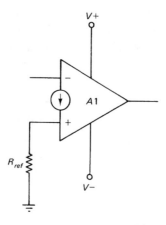

FIGURE 8-15 Using dual polarity power supplies for the CDA.

For best results a multiple CDA integrated circuit such as the LM-3900 should be used for this application. Such a circuit would allow the drift to be controlled because the two halves would both drift at the same rate (they share a common thermal environment).

8-4.8 Using bipolar DC power supplies

The current difference amplifier is designed primarily for single-polarity power supply circuits. In most cases, the CDA will operate with a $V+$ DC power supply in which one side is grounded. We can, however, operate the CDA in a circuit with a bipolar DC power supply using a circuit such as Fig. 8-15. The reference resistor, R_{ref}, is connected from the noninverting input to ground. The $V-$ and $V+$ power supplies are each ground referenced and of equal potential. Thus, the 5 to 100 μA bias current is found from $(V+)/R_{ref}$.

8-5 SUMMARY

1. Operational transconductance amplifiers (OTA) have a transfer function that relates an output current to an input voltage. Thus, the units of gain in the OTA are units of transconductance (siemens and the sub-units). Examples of OTA devices include the CA-3080 and CA-3060 (which is a quad OTA).

2. An OTA can be used as a voltage amplifier by virtue of the voltage drop caused by the output current in an external load resistance.

3. Gain on the OTA is programmed externally with a bias current applied to a special pin on the IC.

4. Current difference amplifiers (CDA), also known as Norton amplifiers, use current inputs but nonetheless have a transfer equation approximately equal to the ratio of input and feedback resistances.

5. The CDA is used especially where only one power supply is available, although bipolar DC power supplies can be accommodated as well.

6. An example of the CDA is the LM-3900 quad Norton amplifier offered by National Semiconductor.

8-6 RECAPITULATION

Now return to the objectives and Pre-quiz questions at the beginning of the chapter and see how well you can answer them. If you cannot answer certain questions, place a check mark to each and review the appropriate parts of the text. Next, try to answer the questions and work the problems below, using the same procedure.

8-7 STUDENT EXERCISES

1. Design a gain-of-200 AC-coupled amplifier based on the LM-3900 CDA. Make the input impedance at least 10 kohms, and the -3 dB low-end frequency response equal to 10 Hz or less. Assume a DC power supply of ± 12 VDC.

2. Design a gain-of-10 AC differential amplifier based on the CDA. Assume an input impedance of 10 kohms or more, and DC power supplies of ± 15 VDC.

3. Design a gain-of-50 voltage amplifier based on the CA-3080 operational transconductance amplifier.

4. Design an analog voltage multiplier based on an OTA.

8-8 QUESTIONS AND PROBLEMS

1. Write the generalized transfer equation for an operational transconductance amplifier (OTA).

2. What bias current is required on a CA-3080 OTA to create a transconductance of 30 mmhos?

3. Calculate the output resistance (R_o) of a CA-3080 if the bias current is 0.75 mA.

4. An OTA has an output load resistance of 10 kohms. When connected as a voltage amplifier an output potential of 5 volts is found across this resistance. Calculate the output current.

5. Draw a circuit diagram that will convert a current-output OTA to a voltage amplifier configuration with a low output impedance.

6. In a current difference amplifier (CDA) the base bias currents of the input transistors is 0.25 mA. Calculate the approximate dynamic resistance (r) at room temperature.

7. Using the dynamic resistance calculated in the above problem, calculate the gain of a CDA voltage amplifier in which the feedback resistance is 100 kohms, and the input resistance is 12 kohms.

8. What value of reference resistor (R_{ref}) is needed to create a 50 µA reference current from a 12 VDC power supply?

9. A CDA is connected to ±15 VDC power supplies. Assuming a 1.5 megohm resistor (R_{ref}) between the 15 VDC supply and the input bias terminal, and a feedback resistance of 120 kohms, calculate the DC offset potential at room temperature when no input signal is present.

10. In a noninverting follower CDA circuit the feedback resistance is 220 kohms, an input resistance of 15 kohms, and a reference current of 100 μA. Calculate the voltage gain.

11. Find the gain of a 'supergain' CDA circuit (Fig. 8-6) if $R1 = 12$ kohms, $R2 = 120$ kohms, $R3 = 10$ kohms, and $R4 = 120$ kohms.

12. Calculate the transconductance of an amplifier in which an output current change of 2 mA is created by an input voltage change of 2 volts.

13. Calculate the output resistance of an OTA in which the bias current of 200 μA is flowing.

14. Draw the circuit for an inverting follower amplifier based on an OTA.

15. What bias current for a CA-3080 OTA will create a transconductance of 40 millisiemens?

16. What bias current is required in a CA-3080 if the transconductance is 25 millisiemens?

17. Design a CDA circuit in which the voltage gain is 230.

18. A CDA inverting follower amplifier has a feedback resistor of 12 kohms, an input resistance of 1 kohm and a current mirror resistance of 100 ohms.

19. A CDA reference resistor has a value of 1 megohm, and is connected to the +15 VDC power supply; the feedback resistor is 220 kohms. Calculate the zero-signal DC output potential that will be produced.

20. Draw the circuit for a CDA amplifier that has a gain of 200, and a lower −3 dB frequency response of 10 Hz. Select the DC power supply potentials and label the components for value.

21. Find the gain of a supergain CDA circuit (Fig. 8-6) if $R1 = 15$ kohms, $R2 = 150$ kohms, and $R4 = 100$ kohms.

22. Find the gain of a supergain CDA circuit (Fig. 8-6) if $R1 = 22$ kohms, $R2 = 120$ kohms and $R4 = 120$ kohms.

23. Draw the circuit for a CDA line driver amplifier with a differential 600-ohm output.

CHAPTER 9

High frequency, VHF, UHF and microwave linear IC devices

OBJECTIVES

1. Understand the problems inherent in high frequency IC applications.
2. Learn the properties of high frequency linear IC devices.
3. Learn the properties of broadband devices.
4. Know the properties of microwave linear IC devices.

9-1 PRE-QUIZ

These questions test your prior knowledge of the material in this chapter. Try answering them before you read the chapter. Look for the answers (especially those you answered incorrectly) as you read the text. After you have finished studying the chapter try answering these questions again, and those at the end of the chapter (see Section 9-11).

1. A Darlington amplifier is constructed with a single collector to base feedback resistor, and an unbypassed emitter resistor. Calculate the value of the emitter resistor if the feedback resistor is 470 ohms, and the input/output impedances must be 50 ohms.

2. Two microwave NPN transistors are connected into a Darlington configuration. Calculate the overall *beta* if both are rated at $\beta = 120$.

3. Calculate the width required of a stripline to match an impedance of 50 ohms if the stripline is 0.125 inches above a groundplane on a printed wiring board that has a dielectric constant of 3.45.

4. Draw a simple schematic of two MIC amplifiers in parallel. Describe the effect on: (a) *driving power requirements*, (b) *output power delivered*, (c) *overall power* and *voltage gain*, and (d) *1 dB compression point*.

9-2 HIGH FREQUENCY LINEAR INTEGRATED CIRCUITS _____

High frequency linear solid-state amplifiers (i.e. those operating from near-DC to VHF, UHF or microwave frequencies) have proven very difficult to design and build with consistent performance across a wide passband. Such amplifiers often have gain irregularities such as 'suck-outs' or peaks in the frequency response. Others suffer large variations of input and output impedance over the frequency range. Still others suffer spurious oscillation at certain frequencies within the passband. Barkhausen's criteria for oscillation requires: (a) loop gain of unity or more, and (b) 360° (in-phase) feedback at the frequency of oscillation. At some frequency, the second of these criteria may be met by adding the normal 180° phase shift inherent in an inverting amplifier to propagation phase shift due to stray *RLC* components in the circuit. The result is oscillation at the frequency where the propagation phase shift is 180°. In addition, in the past only a few applications required such amplifiers. Consequently, such amplifiers were either very expensive, or didn't work nearly as well as claimed. Today, one can buy linear IC devices that work well into the microwave region.

9-3 WHAT ARE HMICS AND MMICS? _____

MMICS are tiny 'gain block' monolithic integrated circuits that operate from DC or near-DC to a frequency in the microwave region. HMICs, on the other hand, are hybrid devices that combine discrete and monolithic technology. One product (Signetics NE-5205) offers up to +20 dB of gain from DC to 0.6 GHz, while another low-cost device (Minicircuits Laboratories, Inc. MAR-x) offers +20 dB of gain over the range DC to 2 GHz depending upon model. Other devices from other manufacturers are also offered, and some produce gains to +30 dB, and frequencies to 18 GHz. Such devices are unique in that they present input and output impedances that are a good match to the 50 or 75 ohms normally used as system impedances in RF circuits.

Monolithic integrated circuits. These devices are formed through photoetching and diffusion processes on a substrate of silicon or some other semiconductor material. Both active devices (such as transistors and diodes) and some passive devices can be formed in this manner. Passive components such as on-chip capacitors and resistors can be formed using various thin and thick film technologies. In the MMIC device, interconnections are made on the chip via built-in planar transmission lines.

Hybrids. Even though physically they often resemble larger monolithic integrated circuits from the outside, hybrids are actually a level closer to regular discrete circuit construction than ICs. Passive components and planar transmission lines are laid down on a glass, ceramic or other insulating substrate by vacuum deposition or some other production method. Transistors and unpackaged monolithic 'chip dies' are cemented to the substrate and then connected to the substrate circuitry via mil-sized gold or aluminum bonding wires.

Because the material in this chapter could apply to either HMIC or MMIC devices, the convention herein shall be to refer to all devices in either subfamily as 'microwave integrated circuits' (MIC), unless otherwise specified.

Three things specifically characterize the MIC device. First, is simplicity. As you will see in the circuits discussed below, the MIC device usually has only input, output, ground and power supply connections. Other wideband IC devices often have up to 16 pins, most of which must be either biased or capacitor bypassed. The second feature of the MIC is the very wide frequency range (DC–GHz) of the devices, while the third is the constant input and output impedance over several octaves of frequency.

Although not universally the case, MICs tend to be unconditionally stable because of a combination of series and shunt negative feedback internal to the device. The input and output impedances of the typical MIC device are a close match to either 50 or 75 ohms, so it is possible for a MIC amplifier to operate without any external impedance matching schemes... a factor that makes it easier to broadband than would be the case if tuning was used. A typical MIC device generally produces a standing wave ratio (SWR) of less than 2:1 at all frequencies within the passband, provided that it is connected to the design system impedance (e.g. 50 ohms). Although the MIC is not usually regarded as a low noise amplifier (LNA), but can produce noise figures (NF) in the 5 to 8 dB range for frequencies up to several gigahertz. Some MICs are LNAs, however, and produce noise figures from 2.5 to 4 dB. The number of LNA MICs that are available on the market will increase in the near future.

Narrowband and passband amplifiers can be built using wideband MICs. A narrowband amplifier is a special case of a passband amplifier, and is typically tuned to a single frequency. An example is the 70 MHz IF amplifier used in microwave receivers. Because of input or output tuning, such an amplifier will respond only to signals near the 70 MHz center frequency.

9-4   VERY WIDEBAND AMPLIFIERS

Engineering wideband amplifiers, such as those used in MIC devices, seems simple but has traditionally caused a lot of difficulty for designers. Figure 9-1A shows the most fundamental form of MIC amplifier circuit; it is a common emitter NPN bipolar transistor amplifier. Because of the high frequency operation of these devices, the amplifier in MICs are usually made of a material such as gallium arsenide (GaAs).

In Fig. 9-1A, the emitter resistor (R_e) is unbypassed, so introduces a small amount of negative feedback into the circuit. Resistor R_e forms *series feedback* for transistor Q1. The *parallel feedback* in this circuit is provided by collector–base bias resistor R_f. Typical values for R_f are in the 500 ohm range, and for R_e in the 4 to 6 ohms range. In general, the designer tries to keep the ratio R_f/R_e high in order to obtain higher gain, higher output power compression points, and lower noise figures. The input and output impedances (R_o) are equal, and defined by the patented equation:

$$R_o = \sqrt{R_f \times R_e} \qquad (9\text{-}1)$$

where:

R_o is the output impedance in ohms (Ω)
R_f is the shunt feedback resistance in ohms (Ω)
R_e is the series feedback resistance in ohms (Ω)

A more common form of MIC amplifier circuit is shown in Fig. 9-1B. Although based on the *Darlington amplifier* circuit, this amplifier still has

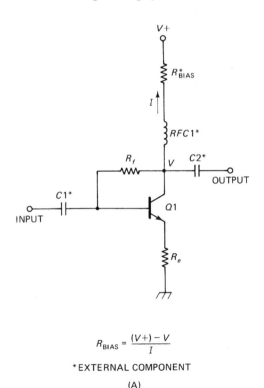

$$R_{BIAS} = \frac{(V+) - V}{I}$$

*EXTERNAL COMPONENT

(A)

FIGURE 9-1   (A) Basic MIC circuit using single transistor; (B) constant input and output impedance MIC amplifier; (C) circuit using MIC device.

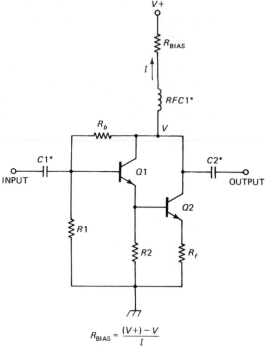

$$R_{BIAS} = \frac{(V+) - V}{I}$$

*EXTERNAL COMPONENTS

(B)

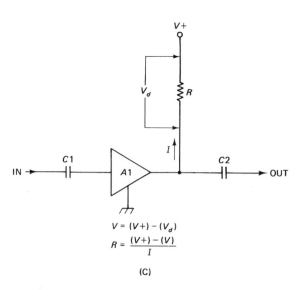

$$V = (V+) - (V_d)$$

$$R = \frac{(V+) - (V)}{I}$$

(C)

FIGURE 9-1 (*continued*)

the same sort of series and shunt feedback resistors (R_e and R_f) as the previous circuit. All resistors except R_{bias} are internal to the MIC device.

A Darlington amplifier, also called a *Darlington pair* or *superbeta transistor*, consists of a pair of bipolar transistors ($Q1$ and $Q2$) connected such that $Q1$ is an emitter follower driving the base of $Q2$, and with both collectors connected in parallel. The Darlington connection permits both transistors to be treated as if they were a single transistor with higher than normal input impedance and a beta gain (β) equal to the product of the individual *beta* gains. For the Darlington amplifier, therefore, the beta (β, or H_{fe}) is:

$$\beta_o = \beta_{Q1} \times \beta_{Q2} \tag{9-2}$$

where:

β_o is the total beta gain of the $Q1/Q2$ Darlington pair
β_{Q1} is the beta gain of $Q1$
β_{Q2} is the beta gain of $Q2$

You should be able to see two facts. First, the *beta* gain is very high for a Darlington amplifier that is made with transistors of relatively modest individual beta figures. Second, the *beta* of a Darlington amplifier made with identical transistors is the square of the common beta rating.

External components. Figures 9-1A and 9-1B show several components that are usually external to the MIC device. The bias resistor (R_{bias}) is sometimes internal, although on most MIC devices it is external. RF choke $RFC1$ is in series with the bias resistor, and is used to enhance operation at the higher frequencies; $RFC1$ is considered optional by some MIC manufacturers. The reactance of the RF choke is in series with the bias resistance, and increases with frequency according to the $2\pi FL$ rule. Thus, the transistor sees a higher impedance load at the upper end of the passband than at the lower end. Use of $RFC1$ as a 'peaking coil' thus helps overcome the adverse effect of stray circuit capacitance that ordinarily causes a similar decreasing frequency-dependent characteristic. A general rule of thumb is to make the combination of R_{bias} and X_{RFC1} form an impedance of at least 500 ohms at the lowest frequency of operation. The gain of the amplifier may drop about 1 dB if $RFC1$ is deleted. This effect is caused by the bias resistance shunting the output impedance of the amplifier.

The capacitors are used to block DC potentials in the circuit. They prevent intra-circuit potentials from affecting other circuits, as well as preventing potentials in other circuits from affecting MIC operation. More will be said about these capacitors later, but for now understand that practical capacitors are not ideal; real capacitors are actually complex *RLC* circuits. While the *L* and *R* components are negligible at low frequencies, they are substantial in the VHF through microwave region. In addition, the LC characteristic forms a self-resonance that can either reduce (suck-out) or enhance (peak-up) gain at specific frequencies. The result is either an uneven frequency response characteristic or spurious oscillations.

9-5 GENERIC MMIC AMPLIFIER

Figure 9-1C shows a 'generic' circuit representing MIC amplifiers in general. As you will see when we look at an actual product, this circuit is nearly complete. The MIC device usually has only input, output, ground and power connections; some models don't have separate DC power input. There is no DC biasing, no bypassing (except at the DC power line) and no seemingly 'useless' pins on the package. MICs tend to use either microstrip packages like UHF/microwave small-signal transistor packages, or small versions of the miniDIP or metallic IC packages. Some HMICs are packaged in larger transistor-like cases, while others are packaged in special hybrid packages.

The bias resistor (R_{bias}) connected to either the DC power supply terminal (if any) or the output terminal. It must be set to a value that limits the current to the device, and drops the supply voltage to a safe value. MIC devices typically require a low DC voltage (4–7 VDC), and a maximum current of about 15 to 25 mA depending upon type. There may also be an optimum current of operation for a specific device. For example, one device advertises that it will operate over a range 2 to 22 mA, but that the optimum design current is 15 mA. The value of resistor needed for R_{bias} is found from Ohm's law:

$$R_{bias} = \frac{(V+) - V}{I_{bias}} \qquad (9\text{-}3)$$

where:

R_{bias} is in ohms (Ω)

$V+$ is the DC power supply potential in volts (V)

V is the rated MIC device operating potential in volts (V)

I_{bias} is the operating current in amperes (A)

The construction of amplifiers based on MIC devices must follow microwave practices. This requirement means short, wide, low-inductance leads made of printed circuit foil, and stripline construction. Interconnection conductors tend to behave like transmission lines at microwave frequencies, so must be treated as such. In addition, capacitors should be capable of passing the frequencies involved, yet have as little inductance as possible. In some cases, the series inductance of common capacitors forms a resonance at some frequency within the passband of the MMIC device. These resonant circuits can sometimes be detuned by placing a small ferrite bead on the capacitor lead. Microwave 'chip' capacitors are used for ordinary bypassing.

MIC technology is currently able to provide very low-cost microwave amplifiers with moderate gain and noise figure specifications, and better

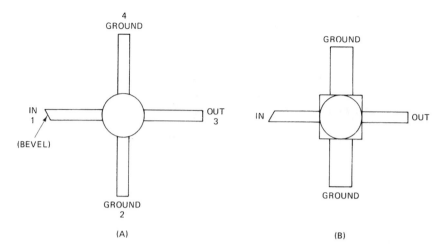

FIGURE 9-2   MIC device package styles.

performance is available at higher cost; the future holds promise of even greater advances. Manufacturers have extended MIC operation to 33 GHz, and reduced noise figures substantially. In addition, it is possible to build the entire front-end of a microwave receiver into a single HMIC or MMIC, including the RF amplifier, mixer and local oscillator stages.

Although MIC devices are available in a variety of package styles, those shown in Fig. 9-2 are typical of many. Because of the very high frequency operation of these devices, MICs are packaged in *stripline* transistor-like cases. The low-inductance leads for these packages are essential in UHF and microwave applications.

9-5.1   Attenuators in amplifier circuits?

It is common practice to place attenuator pads in series with the input and output signal paths of microwave circuits in order to 'swamp-out' impedance variations that adversely affect circuits which ordinarily require either impedance matching or a constant impedance. Especially when dealing with devices such as *LC* filters (low-pass, high-pass, bandpass), VHF/UHF amplifiers, matching networks, and MIC devices, it is useful to insert 1 dB, 2 dB or 3 dB resistor attenuator pads in the input and output lines. The characteristics of many RF circuits depend on seeing the design impedance at input and output terminals. With the attenuator pad (see Fig. 9-3) in the line, source and load impedance changes don't affect the circuit nearly as much.

The attenuator tactic is also sometimes useful when confronted with seemingly unstable very wideband amplifiers. Insert a 1 dB pad in series with both input and output lines of the unstable amplifier. This tactic will cost about 2 dB of voltage gain, but often cures instabilities that arise out of frequency-dependent load or source impedance changes.

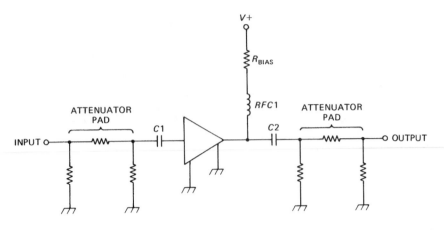

FIGURE 9-3   Use of input and output attenuator pads to stabilize circuit against source and load impedance variation.

9-5.2   Mini-circuits® MAR-x series devices: a MIC example

The MiniCircuits Laboratories, Inc. MAR-x series of MIC devices offers gains from +13 dB to +20 dB, and top-end frequency response of either 1 GHz or 2 GHz depending upon type. The package used for the MAR-x device (Fig. 9-2A) is similar to the case used for UHF and microwave transistors. Pin No. 1 (RF input) is marked by a color dot and a bevel. The usual circuit for the MAR-x series devices is shown in Fig. 9-4. The MAR-x device requires a voltage of +5 Vdc on the output terminal, and must derive this potential from a DC supply of greater than +7 Vdc.

The RF choke ($RFC1$) is called 'optional' in the engineering literature on the MAR-x, but it is recommended for applications where a substantial portion of the total bandpass capability of the device is used. The choke tends to pre-emphasize the higher frequencies, and thereby overcomes the de-emphasis normally caused by circuit capacitance; in traditional video amplifier terminology that coil is called a 'peaking coil' because of this action, i.e. it peaks-up the higher frequencies.

It is necessary to select a resistor for the DC power supply connection. The MAR-x device wants to see +5 Vdc, at a current not to exceed 20 mA. In addition, $V+$ must be greater than +7 volts. Thus, we need to calculate a dropping resistor (R_d) of

$$R_d = \frac{(V+) - 5 \text{ Vdc}}{I} \tag{9-4}$$

where:
   R_d is in ohms (Ω)
   I is in amperes (A)

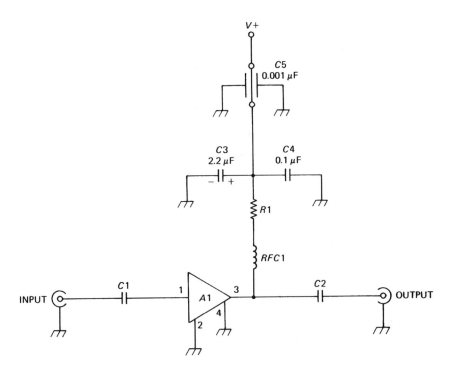

FIGURE 9-4   Basic circuit for the MAR-x series MIC devices.

In an amplifier designed for +12 VDC operation select a trial bias current of 15 mA (i.e. 0.015 A). The resistor value calculated is 467 ohms (a 470 ohm, 5%, film resistor will prove satisfactory). As recommended above, a 1 dB attenuator pad is inserted in the input and output lines. The V+ is supplied to this chip through the output terminal.

9-6   CASCADE MIC AMPLIFIERS

MIC devices can be connected in cascade (Fig. 9-5) to provide greater gain than is available from only a single device, although a few precautions must be observed. It must be recognized, for example, that MICs possess a substantial amount of gain from frequencies near DC to well into the microwave region. In all cascade amplifiers attention must be paid to preventing feedback from stage to stage. Two factors must be addressed. First, as always, is component layout. The output and input circuitry external to the MIC must be physically separated in order to prevent coupling feedback. Second, it is necessary to decouple the DC power supply lines that feed two or more stages. Signals carried on the DC power line can easily couple into one or more stages, resulting in unwanted feedback.

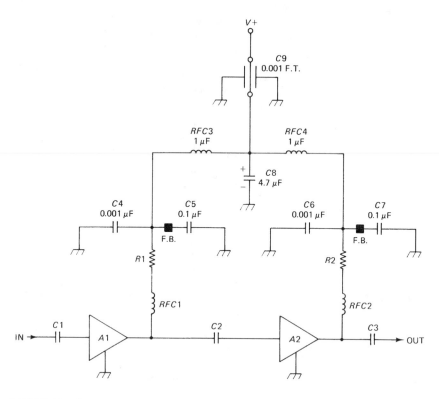

FIGURE 9-5   Cascade MAR-x MIC devices.

Figure 9-5 shows a method for decoupling the DC power line of a two-stage MIC amplifier. In lower frequency cascade amplifiers the $V+$ ends of resistors $R1$ and $R2$ would normally be joined together and connected to the DC power supply. Only a single capacitor would be needed at that junction to ensure adequate decoupling between stages. But as the operating frequency increases the situation becomes more complex, in part due to the non-ideal nature of practical components. For example, in an audio amplifier the same aluminum electrolytic capacitors that are used in the power supply ripple filter may provide sufficient decoupling. At RF frequencies, however, electrolytic capacitors are almost useless as capacitors (they act more like resistors at those frequencies).

The decoupling system in Fig. 9-5 consists of RF chokes $RFC3$ and $RFC4$, and capacitors $C4$ through $C9$. The RF chokes help prevent high frequency AC signals from traveling along the DC power line. These chokes are selected for a high reactance at VHF through microwave frequencies, while having a low DC resistance. For example, a one microhenry (1 μH) RF choke might have only a few milliohms of DC resistance, but (by $2\pi FL$) has a reactance of more than 3000 ohms at 500 MHz. It is important that

$RFC3$ and $RFC4$ be mounted so as to minimize mutual inductance due to interaction of their respective magnetic fields.

Capacitors $C4$ through $C9$ are used for bypassing signals to ground. It should be noted that a wide range of values, and several types of capacitor, are used in this circuit. Each has its own purpose. Capacitor $C8$, for example, is an electrolytic type and is used to decouple very low frequency signals (i.e. those up to several hundred kilohertz). Because $C8$ is ineffective at higher frequencies it is shunted by capacitor $C9$, shown in Fig. 9-5 as a feedthrough capacitor. Such a capacitor is usually mounted on the shielded enclosure housing the amplifier. Capacitors $C5$ and $C7$ are used to bypass signals in the HF region. Because these capacitors are likely to exhibit substantial series inductance, they may form undesirable resonances within the amplifier passband. Ferrite beads (FB) are sometimes installed on each capacitor lead in order to detune capacitor self-resonances. Like $C9$, capacitors $C4$ and $C6$ are used to decouple signals in the VHF-and-up region. These capacitors must be of microwave chip construction, or they may prove ineffective above 200 MHz or so.

9-6.1 Gain in cascade amplifiers

In low frequency amplifiers we might reasonably expect the composite gain of a cascade amplifier to be the product of the individual stage gains:

$$G = G1 \times G2 \times G3 \times \cdots \times G_n \qquad (9\text{-}5)$$

While that reasoning is valid for low frequency voltage amplifiers, it fails for RF amplifiers (especially in the microwave region) where input–output standing wave ratio (SWR) becomes significant. In fact, the gain of any RF amplifier cannot be accurately measured if the SWR is greater than about 1.15:1. SWR is a measure of the RF signal power that is absorbed by a load versus the power that is reflected back towards the source. A discussion of SWR (and VSWR) is beyond the scope of this text. It is sufficient to note that SWR can be calculated by taking the ratio of source impedance (R_o) and load impedance (Z_L).

There are several ways in which SWR can be greater than 1:1, and all of them involve an impedance mismatch. For example, the amplifier may have an input or output resistance other than the specified value. This situation can arise due to design errors or manufacturing tolerances. Another source of mismatch is the source and load impedances. If these impedances are not exactly the same as the amplifier input or output impedance, respectively, then a mismatch will occur.

An impedance mismatch at either input or output of the single-stage amplifier will result in a gain mismatch loss (ML) of:

$$ML = -10 \log \left[1 - \left[\frac{SWR - 1}{SWR + 1} \right]^2 \right] \qquad (9\text{-}6)$$

where:

ML is the mismatch loss in decibels (dB)

SWR is the standing wave ratio (dimensionless)

In a cascade amplifier we have the distinct possibility of an impedance mismatch, hence an SWR, at more than one point in the circuit. An example (refer again to Fig. 9-5) might be where neither the output impedance (R_o) of the driving amplifier ($A1$) nor the input impedance (R_i) of the driven amplifier ($A2$) are matched to the 50 ohm (Z_o) microstrip transmission line that interconnects the two stages. Thus, R_o/Z_o or its inverse forms one SWR, while R_i/Z_o or its inverse forms the other. For a two-stage cascade amplifier the mismatch loss is:

$$ML = -20\log\left[1 \pm \left[\frac{SWR1 - 1}{SWR1 + 1}\right]\left[\frac{SWR2 - 1}{SWR2 + 1}\right]\right] \qquad (9\text{-}7)$$

EXAMPLE 9-1

An amplifier ($A1$) with a 25 ohm output resistance drives a 50 ohm stripline transmission line (Fig. 9-6). The other end of the stripline is connected to the input of another amplifier ($A2$) in which $R_i = 100$ ohms; (a) calculate the maximum and minimum gain loss for this system, and (b) calculate the range of system gain if $G1 = 6$ dB and $G2 = 10$ dB.

Solution

$$SWR1 = Z_o/R_o = 50/25 = 2{:}1$$

$$SWR2 = R_i/Z_o = 100/50 = 2{:}1$$

$$ML = -20\log\left[1 \pm \left[\frac{SWR1 - 1}{SWR1 + 1}\right]\left[\frac{SWR2 - 1}{SWR2 + 1}\right]\right]$$

$$ML = -20\log\left[1 + \left[\frac{2 - 1}{2 + 1}\right]\left[\frac{2 - 1}{2 + 1}\right]\right]$$

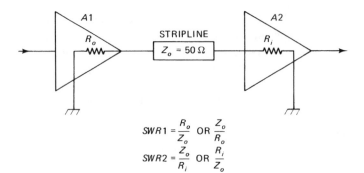

FIGURE 9-6   Use of stripline section between stages.

$$ML1 = 20 \log[1 + (1/3)(1/3)] \text{ dB}$$

$$ML1 = 20 \log[1 + 0.11] \text{ dB}$$

$$ML1 = 20 \log[1.11] \text{ dB}$$

$$ML1 = (20)(0.045) = 0.91 \text{ dB}$$

$$ML2 = 20 \log(1 - 0.11) \text{ dB}$$

$$ML2 = 20 \log(0.89) \text{ dB}$$

$$ML2 = (20)(-0.051) \text{ dB}$$

$$ML2 = -1.02 \text{ dB}$$

Thus,

$$ML1 = +1.11 \text{ dB}$$

$$ML2 = -1.02 \text{ dB}$$

Without the SWR, gain in decibels is $G = G1 + G2 = (6 \text{ dB}) + (10 \text{ dB}) = 16$ dB. With SWR considered we find that $G = G1 + G2 \pm ML$. So,

$$G_a = G1 + G2 + ML1$$

$$G_a = (6 \text{ dB}) + (10 \text{ dB}) + (1.11 \text{ dB})$$

$$G_a = 17.11 \text{ dB}$$

and

$$G_b = G1 + G2 + ML2$$

$$G_b = (6 \text{ dB}) + (10 \text{ dB}) + (-1.02 \text{ dB})$$

$$G_b = 14.98 \text{ dB} \qquad \blacksquare$$

The mismatch loss can vary from a negative loss resulting in less system gain (G_b), to a 'positive loss' (which is actually a gain in its own right) resulting in greater system gain (G_a). The reason for this apparent paradox is that it is possible for a mismatched impedance to be connected to its complex conjugate impedance.

9-7 COMBINING MIC AMPLIFIERS IN PARALLEL _____

Figures 9-7 through 9-10 show several MIC applications involving combinations of two or more MIC devices. Perhaps the simplest of these is the configuration of Fig. 9-7. Although each input must have its own DC blocking capacitor to protect the MIC device's internal DC bias network, the outputs of two or more MICs may be connected in parallel and share a common power supply connection and output coupling capacitor.

Several advantages are realized with the circuit of Fig. 9-7. First, the power output increases even though total system gain (P_o/P_m) remains the same. As

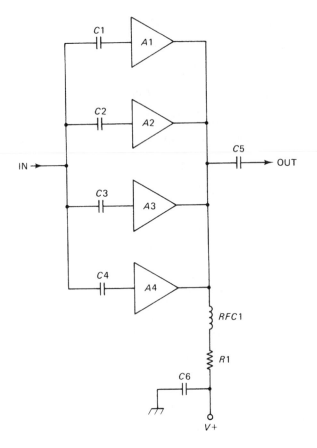

FIGURE 9-7   Parallel connection to increase power gain of MIC devices.

a consequence, however, drive power requirements also increase. The output power increases 3 dB when two MICs are connected in parallel, and 6 dB when four are connected (as shown). The 1 dB output power compression point also increases in parallel amplifiers in the same manner: 3 dB for two amplifiers and 6 dB for four amplifiers in parallel.

The input impedance of a parallel combination of MIC devices reduces to R_i/N, where N is the number of MIC devices in parallel. In the circuit shown in Fig. 9-7, the input impedance would be $R_i/4$, or 12.5 ohms if the MICs are designed for 50 ohm service. Because 50/12.5 represents a 4:1 SWR, some form of input impedance matching must be used. Such a matching network can be either broadbanded or frequency specific as the need dictates. Figures 9-8 and 9-9 show methods for accomplishing impedance matching.

The method shown in Fig. 9-8 is used at operating frequencies up to about 100 MHz, and is based on broadband ferrite toroidal RF transformers. These

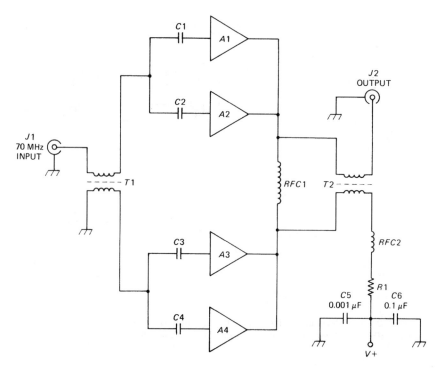

FIGURE 9-8   70 MHz MAR-x MIC amplifier using hybrid transformers.

transformers dominate the frequency response of the system because they are less broadbanded than the usual MIC device. This type of circuit can be used as a gain block in microwave receiver IF amplifiers (which are frequently in the 70 MHz region), or in the exciter section of master oscillator power amplifier (MOPA) transmitters.

Another method is more useful in the UHF and microwave regions. In Fig. 9-9 we see several forms of the *Wilkinson power divider* circuit. An *LC* network version is shown in Fig. 9-9A for comparison, although coaxial (Fig. 9-9B) or stripline (Figs 9-9C and 9-9D) transmission line versions are used in microwave applications. The *LC* version is used to frequencies of 150 MHz. This circuit is bidirectional, so can be used as either a *power splitter* or *power divider*. RF power applied to port C is divided equally between port A and port B. Alternatively, power applied to ports A/B are summed together and appear at port C. The component values are found from the following relationships:

$$R = 2Z_0 \tag{9-8}$$

$$L = \frac{70.7}{2\pi F_0} \tag{9-9}$$

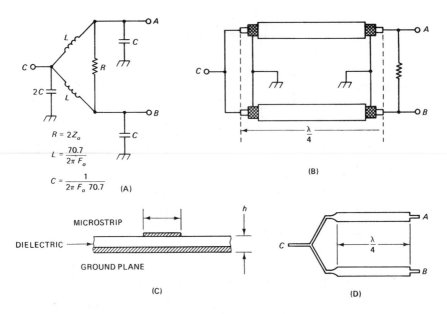

FIGURE 9-9 (A) Wilkinson divider circuit; (B) transmission line (coax) implementation of Wilkinson divider; (C) section view of stripline; (D) stripline version of the Wilkinson divider.

$$C = \frac{70.7}{2\pi 70.7 F_0} \tag{9-10}$$

where:
  R is in ohms (Ω)
  L is in henrys (H)
  C is in farads (F)
  F_0 is the frequency in hertz (Hz)

Figure 9-9B shows a coaxial cable version of the Wilkinson divider that can be used at frequencies up to about 2 GHz. The lower frequency limit is set by practicality because the transmission line segments become too long to be handled easily. The upper frequency limit is set by the practicality of handling very short lines, and by the dielectric losses which are frequency dependent. The transmission line segments are each quarter wavelength; their length is found from:

$$L = \frac{2952V}{F} \tag{9-11}$$

where:
  L is the physical length of the line in inches (in).
  F is the frequency in megahertz (MHz)
  V is the *velocity factor* of the transmission line $(0-1)$

An impedance transformation can take place across a quarter wavelength transmission line if the line has a different impedance than the source or load impedances being matched. Such an impedance matching system is often called a 'Q-section'. The required characteristic impedance for the transmission line is found from:

$$Z'_0 = \sqrt{Z_L Z_0} \qquad (9\text{-}12)$$

where:

Z'_0 is the characteristic impedance of the quarter wavelength section in ohms (Ω)

Z_L is the load impedance in ohms (Ω)

Z_0 is the system impedance in ohms (Ω), (e.g. 50 ohms)

In the case of parallel MIC devices, the nominal impedance at port C of the Wilkinson divider is one-half of the reflected impedance of the two transmission lines. For example, if the two lines are each 50 ohm transmission lines, then the impedance at port C is 50/2 ohms, or 25 ohms. Similarly, if the impedance of the load, i.e. the reflected impedance, is transformed to some other value, then port C sees the parallel combination of the two transformed impedances. In the case of a parallel MIC amplifier we might have two devices with 50 ohms input impedance each. Placing these devices in parallel halves the impedance to 25 ohms, which forms a 2:1 SWR with a 50 ohm system impedance. But if the quarter wavelength transmission line transforms the 50 ohm input impedance of each device to 100 ohms, then the port C impedance is 100/2, or 50 ohms... which is correct.

At the upper end of the UHF spectrum, and in the microwave spectrum, it may be better to use a stripline transmission line instead of coaxial cable. A stripline (see Fig. 9-9C) is formed on a printed circuit board. The board must be double-sided so that one side can be used as a ground plane, while the stripline is etched into the other side. The length of the stripline is dependent upon the frequency of operation; either halfwave or quarterwave lines are usually used. The impedance of the stripline is a function of three factors: (1) *stripline width* (*w*), (2) *height of the stripline above the ground plane* (*h*), and (3) *dielectric constant* (*ε*) of the printed circuit material:

$$Z_0 = 377 \frac{h}{w\sqrt{\varepsilon}} \qquad (9\text{-}13)$$

where:

h is the height of the stripline above the groundplane

w is the width of the stripline (*h* and *w* in same units)

Z_0 is the characteristic impedance in ohms

The stripline transmission line is etched into the printed circuit board as in Fig. 9-9D. Stripline methods use the printed wiring board to form conductors, tuned circuits, etc. In general, for operation at VHF and above the

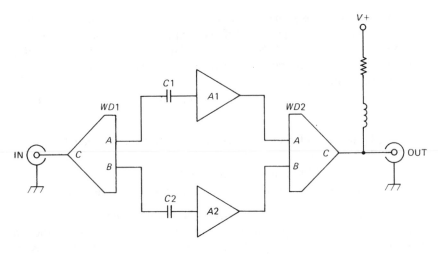

FIGURE 9-10 Using input and output Wilkinson dividers to combine MIC amplifiers.

conductors must be very wide (relative to their simple DC and RF power carrying size requirements), and very short, in order to reduce lead inductance. Certain elements, the actual striplines, are transmission line segments and follow transmission line rules. The printed circuit material must have a *large permitivity* and a *low loss tangent.* For frequencies up to about 3 GHz it is permissible to use ordinary glass-epoxy double-sided board ($\varepsilon = 5$), but for higher frequencies a low-loss material such as Rogers *Duroid* ($\varepsilon = 2.17$) must be used.

When soldering connections in these amplifiers, it is important to use as little solder as possible, and keep the soldered surface smooth and flat as possible. Otherwise, the surface wave on the stripline will be interrupted... and operation suffers.

Figure 9-10 shows a general circuit for a multi-MIC amplifier based on one of the Wilkinson power dividers discussed above. Because the divider can be used as either splitter or combiner, the same type can be used as input ($WD1$) and output ($WD2$) terminations. In the case of the input circuit, port C is connected to the amplifier main input, while ports A/B are connected to the individual inputs of the MIC devices. At the output circuit another divider ($WD2$) is used to combine power output from the two MIC amplifiers, and direct it to a common output connection. Bias is supplied to the MIC amplifiers through a common DC path at port C of the Wilkinson dividers.

9-8 CONCLUSION

The world of the microwave device is no longer 'becoming more important' in the arena of linear IC technology, but rather it is arrived. Microwaves are a crowded portion of the electromagnetic spectrum, and there has been a great increase in the number of MIC devices that are available for that market.

9-9 SUMMARY

1. Microwave integrated circuits (MIC) are monolithic or hybrid gain blocks that have (a) bandwidth from DC or near-DC to the microwave range, (b) constant input and output impedance over at least several octaves of frequency, only the bare essentials for connections (e.g. input, output, ground and sometimes DC power). Some MIC devices are unconditionally stable over the entire frequency range.

2. MICs have all components built-in except for input and output coupling capacitors and an external bias resistor. For some types an external peaking inductor or 'RF choke' is recommended. Narrowband or bandpass amplifiers can be built by using tuning devices or frequency selective filtering in the input and/or output circuits.

3. MIC devices can be connected in cascade in order to increase overall system gain. Care must be taken to ensure that the SWR is low, that is the impedances are matched. Otherwise, matching must be provided or gain errors tolerated.

4. The noise figure in a cascade MIC amplifier is dominated by the noise figure of the input amplifier stage.

5. MICs can be connected in parallel. The output terminals may be directly connected in parallel, but the input terminals of each device must be provided with its own DC blocking capacitor. MICs can also be combined using broadband RF transformers, Wilkinson power dividers/splitters or other methods.

6. Printed circuit boards used for MIC devices must have a high permitivity and a low loss tangent. For frequencies up to 3 GHz glass-epoxy is permissible, but for higher frequencies a low-loss product is required.

9-10 RECAPITULATION

Now return to the objectives and Pre-quiz questions at the beginning of the chapter and see how well you can answer them. If you cannot answer certain questions, place a check mark to each and review the appropriate parts of the text. Next, try to answer the questions and work the problems below, using the same procedure.

9-11 QUESTIONS AND PROBLEMS

1. What condition may result in a microwave amplifier if the loop gain is unity or greater, and the feedback in inadvertently in-phase with the input signal?

2. On-chip interconnections in MIC devices are usually made with _____ transmission lines.

3. What type of semiconductor material is usually used to make the active devices in MIC amplifiers?

4. A MIC amplifier contains a common emitter Darlington amplifier transistor and internal _____ and _____ negative feedback elements.

5. In a common emitter MIC amplifier the _____ ratio must be kept high in order to obtain higher gain, higher output power compression points and lower noise figure.

6. A common emitter amplifier inside a MIC device has a collector–base resistor of 450 ohms, and an emitter resistor of 12.5 ohms. Is this amplifier a good impedance match for 75 ohms cable television amplications?

7. Calculate the input and output impedances of a MIC amplifier in which the series feedback resistor is 5 ohms, and the shunt feedback resistor is 550 ohms.

8. Calculate the gain of the amplifiers in the previous two questions: (a) _____ , and (b) _____ .

9. Two transistors are connected in a Darlington amplifier configuration. $Q1$ has a beta of 100, and $Q2$ has a beta of 75. What is the beta gain of the Darlington pair?

10. An MIC amplifier must be operated over a frequency range of 0.5–2.0 GHz. The DC power circuit consists of an RF choke and a bias resistor in series between $V+$ and the MIC output terminal. If the resistor has a value of 220 ohms, what is the minimum appropriate value of the RF choke required in order to achieve an impedance of at least 500 ohms?

11. A *MCL* MAR-1 must operate from a +12 VDC power supply. This device requires a +5 VDC supply, and has an optimum current of 17 mA. Calculate the value of bias resistor.

12. A MIC amplifier has an optimum current of 22 mA, and requires +7 VDC at the output terminal. Calculate the bias resistor required to operate the device from (a) 12 VDC, and (b) 9 VDC.

13. Because they operate into the microwave region MIC devices are packaged in _____ cases.

14. In order to overcome problems associated with changes of source and load impedance it is sometimes the practice to insert resistive _____ in series with the input and output signal paths of a MIC device.

15. The SWR in a microwave amplifier should be less than _____ in order to make accurate gain measurements.

16. Calculate the mismatch loss in a single-stage MIC amplifier if the output impedance is 50 ohms, and the load impedance is 150 ohms.

17. Calculate the mismatch loss in a single-stage MIC amplifier if the input SWR is 1.75:1.

18. Calculate the mismatch loss in a cascade MIC amplifier if the output impedance of the driver amplifier and the input impedance of the final amplifier are both 50 ohms, but they are interconnected by a 90 ohms transmission line. Remember that there are two values: (a) _____ dB and (b) _____ dB.

19. A cascade MIC amplifier has two SWR mismatches: $SWR1 = 2{:}1$, and $SWR2 = 2.25{:}1$. Calculate both values of mismatch loss.

20. The amplifiers in the previous question have the following gains: $A1 = 16$ dB, and $A2 = 12$ dB. Calculate the range of gain that can be expected from the cascade combination $A1 \times A2$ considering the possible mismatch losses.

21. A MIC amplifier has two stages $A1$ and $A2$. $A1$ has a gain of 20 dB and a noise figure of 3 dB; $A2$ has a gain of 10 dB and a noise figure of 4.5 dB. Calculate the noise figure of the system (a) when $A1$ is the input amplifier and $A2$ is the output amplifier, and (b) the reverse situation, i.e. when $A2$ is the input amplifier and $A1$ is the output amplifier. What practical conclusions do you draw from comparing these results?

22. Two MIC amplifiers are connected in parallel. What is the total input impedance if each amplifier has an input impedance of 75 ohms.

23. Four MIC amplifiers are connected in parallel. The output compression point is now *raised/lowered* _____ dB.

24. An LC Wilkinson power divider must be used to connect two MIC devices in parallel. The system impedance, Z_o, is 50 ohms, and each MIC amplifier has an impedance of 50 ohms. Calculate the values of R, L and C required for this circuit.

25. A quarter wavelength transmission line for 1.296 GHz must be made from *Teflon*® coaxial cable that has a dielectric constant of 0.77. What is the physical length in inches?

26. A 75 ohm load must be transformed to 125 ohms in a quarter-wave Q-section transmission line. Calculate the required characteristic impedance required for this application.

27. A stripline transmission line is being built on a double sided printed circuit board made with low-loss ($\varepsilon = 2.5$) material. If the board is 0.125 inches thick, what width must be stripline be in order to match 70.7 ohms?

CHAPTER 10

IC waveform generators and waveshaping circuits

OBJECTIVES

1. Understand the distinction between relaxation and feedback oscillators.
2. Learn the operation of monostable multivibrators.
3. Learn the operation of astable squarewave, triangle and sawtooth astable multivibrators.
4. Learn the operation of sinewave oscillators.

10-1 PRE-QUIZ

These questions test your prior knowledge of the material in this chapter. Try answering them before you read the chapter. Look for the answers (especially those you answered incorrectly) as you read the text. After you have finished studying the chapter try answering these questions again, and those at the end of the chapter (see Section 10-11).

1. Calculate the RC time constant that will allow a capacitor to charge from -10 Vdc to $+10$ Vdc in 250 ms.
2. A monostable multivibrator (MMV) must produce a 20 ms pulse. Select an appropriate RC time constant and suggest component values.
3. A monostable multivibrator has a timing resistor of 100 kohms, and a timing capacitor of 0.001 μF. What is the duration of the output pulse if the positive feedback resistors are of equal value?
4. Calculate the frequency of an op-amp astable multivibrator in which the positive feedback resistors are equal, and the RC time constant is 0.030 seconds.

10-2 INTRODUCTION TO WAVEFORM GENERATORS _____

Waveform generators are used to produce the large variety of electronic waveforms that are needed in many different circuits. Some waveform generators are sinewave oscillators, even though the word 'oscillator' is also correctly applied to circuits that produce other waveforms. The astable multivibrator may produce square waves, triangle waves or other non-sinusoidal waveforms. Similarly, a *digital clock* is a special case of the multivibrator that is used in digital logic and computer circuits.

The generic term *oscillator* may be used to denote all three cases, including sinewave oscillators, astable multivibrators and digital clocks. The term oscillator can be defined as a *circuit that produces a periodic waveform* (i.e. one that repeats itself). The output waveform can be a sinewave, squarewave, triangle wave, sawtooth wave, pulses or any of several other waveshapes. The important thing is that the waveform is *periodic*.

A class of waveform generator that is not an oscillator is the *monostable multivibrator*, or *one-shot*, circuit. This circuit produces only a single pulse when triggered, so it is not periodic.

There are two basic forms of oscillator circuit: *relaxation oscillators* and *feedback oscillators*. Feedback oscillators use an active device such as an amplifier, and then provide feedback in a manner that produces *regeneration* instead of degeneration. These circuits account for a large number of the oscillators used in practical electronic circuits.

Some relaxation oscillators use one of several available negative resistance devices (e.g. tunnel diode). Such devices operate according to Ohm's law under certain conditions, and different from Ohm's law under other conditions. Other relaxation oscillators use devices that pass little or no current at voltages below some threshold, and pass a large current at voltages above the threshold. Examples of these devices are neon glow-lamps and unijunction transistors (UJT).

There is also a sub-class of oscillators that are based on IC devices such as voltage comparators, operational amplifiers, integrators and so forth. It is these circuits that are discussed in this chapter.

Because the circuits in this chapter are based on the charge and discharge properties of resistor–capacitor networks it is prudent to review the operation of simple *RC* networks.

10-3 REVIEW OF RESISTOR—CAPACITOR (R—C) NETWORKS ____

This section is provided as a brief review of *RC* network DC theory. Consider Fig. 10-1A. Assuming that the initial condition is as shown, switch $S1$ is in position A and is thus open-circuited. There is initially no electrical charge stored in capacitor C (i.e. $V_c = 0$). If switch $S1$ is moved to position B, however, voltage V is applied to the RC network. The capacitor begins to charge with current from the battery, and V_c begins to rise towards V (see curve V_{cb} in Fig. 10-1B). The instantaneous capacitor voltage is found from:

$$V_c = V[1 - e^{-T/RC}] \qquad (10\text{-}1)$$

where:
  V_c is the capacitor charge potential in volts (V)
  V is the applied potential from the source in volts (V)

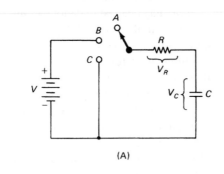

(A)

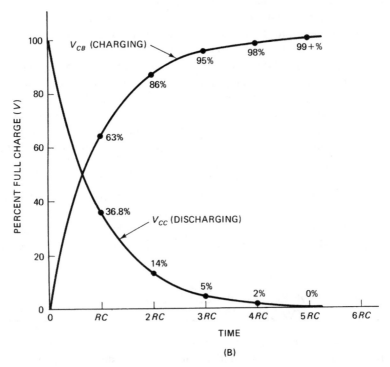

(B)

FIGURE 10-1 (A) RC charge–discharge circuit; (B) charge and discharge curves for RC network; (C) time to charge from initial voltage to a final voltage.

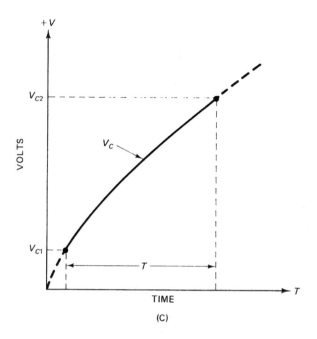

FIGURE 10-1 (continued)

T is the elapsed time after charging begins in seconds (s)
R is the resistance in ohms (Ω)
C is the capacitance in farads (F)

The product RC is called the RC time constant of the network, and is sometimes abbreviated (τ). If R is in *ohms*, and C is in *farads*, then the product RC is in *seconds*. The capacitor voltage rises to approximately 63.2% of the final value after $1RC$, 86% after $2RC$ and >99% after $5RC$. A capacitor in an RC network is considered fully charged (by definition) after five time constants.

If switch $S1$ in Fig. 10-1A is next set to position C, the capacitor will begin to discharge through the resistor. In the discharge condition:

$$V_c = Ve^{-T/RC} \qquad (10\text{-}2)$$

Voltage V_c drops to 36.8% of the full charge level after one time constant ($1RC$), and to very nearly zero after $5RC$.

Next consider Fig. 10-1C. This graph represents a situation commonly encountered in waveform generator circuits. In this graph the capacitor is required to charge from some initial condition (V_{C1}), which may or may not be 0 volts, to a final condition (V_{C2}), which may or may not be the fully charged '$5RC$' point, in a specified time interval, T. The question asked is:

What RC time constant will force V_{C1} to rise to V_{C2} in time T? Assuming that $V_{C1} < V_{C2} < V$:

$$V - V_{C2} = (V - V_{C1})e^{-T/RC} \qquad (10\text{-}3)$$

$$\frac{V - V_{C2}}{V - V_{C1}} = e^{-T/RC} \qquad (10\text{-}4)$$

$$\ln\left[\frac{V - V_{C2}}{V - V_{C1}}\right] = \frac{-T}{RC} \qquad (10\text{-}5)$$

or, rearranging terms:

$$RC = \frac{-T}{\ln\left[\dfrac{V - V_{C2}}{V - V_{C1}}\right]} \qquad (10\text{-}6)$$

EXAMPLE 10-1

An RC network is connected to a +12 Vdc source. What RC product will permit voltage V_c to rise from +1 Vdc to +4 Vdc in 200 ms when $V = +12$ Vdc, $V_{C2} = +4$ Vdc and $V_{C1} = +1$ Vdc?

Solution

$$\ln\left[\frac{V - V_{C2}}{V - V_{C1}}\right] = \frac{-T}{RC}$$

$$RC = \frac{-\left[200 \text{ ms} \times \dfrac{1 \text{ s}}{1000 \text{ ms}}\right]}{\ln\left[\dfrac{12 - 4}{12 - 1}\right]}$$

$$RC = \frac{-0.200 \text{ s}}{\ln\left[\dfrac{8}{11}\right]} = \frac{-0.200 \text{ s}}{\ln(0.727)}$$

$$RC = (-0.200 \text{ s})(-0.319) = 0.0627 \text{ s} \qquad \blacksquare$$

Equation (10-3) can be used to derive the timing or frequency setting equations of many different RC-based waveform generator circuits. The key voltage levels will, most often, be trip points or critical values set by the design of the circuit.

10-4 MONOSTABLE MULTIVIBRATOR CIRCUITS _____

The *monostable multivibrator* (MMV) has two permissible output states (HIGH and LOW), but only one of them is stable. The MMV produces *one output pulse in response to an input trigger signal* (Fig. 10-2). The output pulse (V_o) has a time duration, T, in which the output is in the *quasi-stable*

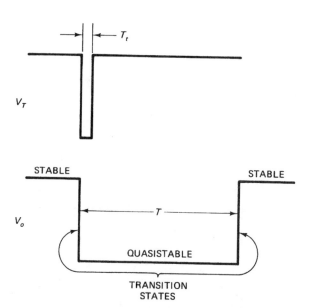

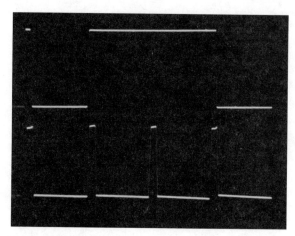

FIGURE 10-2 Trigger and output pulse relationships for a monostable multivibrator (one-shot) circuit.

FIGURE 10-3 Monostable multivibrator does not respond to further input trigger pulses during output pulse duration.

state. The MMV is also known under several alternate names: *one-shot, pulse generator,* and *pulse stretcher.* The latter name derives from the fact that the output pulse duration T is longer than the trigger pulse duration ($T > T_c$).

Monostable multivibrators find a wide variety of applications in electronic circuits. Besides the pulse stretcher mentioned above, the MMV also serves to lock out unwanted pulses. Figure 10-3 shows that the output responds to

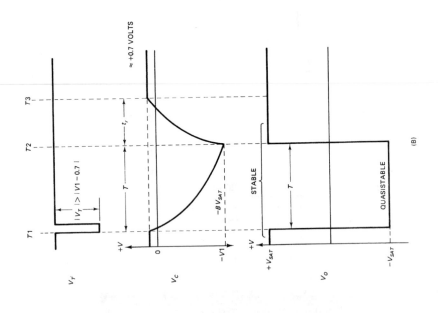

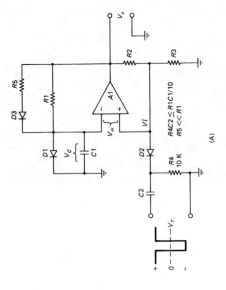

FIGURE 10-4 (A) op-amp monostable multivibrator circuit; (B) timing waveforms.

only the first trigger pulse. The next two pulses occur during the active time, T, so are ignored. Such an MMV is said to be *nonretriggerable*. A common application of this feature is in switch contact 'debouncing'. All mechanical switch contacts bounce a few times on closure, creating a short exponentially decaying run of pulses. If an MMV is triggered by the first pulse from the switch, and if the MMV remains quasi-active long enough for the bouncing to die out, then the MMV output signal becomes the debounced switch closure. The main requirement is that the MMV duration be longer than the switch contact bounce pulse train; 5 ms is generally considered adequate for most switch types.

The range of MMV applications is too broad for detailed discussion here, so only a general set of categories can be presented. These include: *pulse generation, pulse stretching, contact debouncing, pulse signal clean-up, switching*, and *synchronization* of circuit functions (especially digital).

Figure 10-4A shows the circuit for a nonretriggerable monostable multivibrator based on the operational amplifier. This circuit is based on the voltage comparator circuit discussed in Chapter 1. When there is no feedback, the effective voltage gain of an op-amp is its open-loop gain (A_{vol}). When both $-IN$ and $+IN$ are at the same potential, the differential input voltage (V_{id}) is zero, so the output is also zero. But if V_{-IN} does not equal V_{+IN}, the high gain of the amplifier forces the output to either its positive or negative saturation values. If $V_{-IN} > V_{+IN}$, the op-amp sees a positive differential input signal, so the output saturates at $-V_{sat}$. However, if $V_{-IN} < V_{+IN}$, the amplifier sees a negative differential input signal and the output saturates to $+V_{sat}$. The operation of the MMV depends on the relationship of V_{-IN} and V_{+IN}.

There are four states of the monostable multivibrator that must be considered: *stable state, transition state, quasi-stable state*, and *refractory state*.

10-4.1   Stable state

The output voltage V_o is initially at $+V_{sat}$. Capacitor $C1$ will attempt to charge in the positive going direction because $+V_{sat}$ is applied to the $R1C1$ network. But, because of diode $D1$ shunted across $C1$, the voltage across $C1$ is clamped to $+V_{D1}$. For a silicon diode such as the 1N914 or 1N4148 $+V_{D1}$ is about $+0.7$ Vdc. Thus, the inverting input ($-IN$) is held to $+0.7$ Vdc during the stable state. The noninverting input ($+IN$) is biased to a level $V1$, which is:

$$V1 = \frac{R3(+V_{sat})}{R2 + R3} \tag{10-7}$$

or, in the special (but common) case of $R2 = R3$:

$$V1 = \frac{+(V_{sat})}{2} \tag{10-8}$$

The factor $R3/(R2 + R3)$ is often designated by the Greek letter *beta* (β), so:

$$\beta = \frac{R3}{R2 + R3} \qquad (10\text{-}9)$$

Therefore:

$$V1 = \beta(+V_{sat}) \qquad (10\text{-}10)$$

The amplifier ($A1$) sees a differential input voltage (V_{id}) of ($V1 - V_{D1}$), or ($V1 - 0.7$) volts. Using the previous notation:

$$V_{id} = \frac{R3(+V_{sat})}{R2 + R3} \qquad (10\text{-}11)$$

As long as $V1 > V_{D1}$, the amplifier effectively sees a negative DC differential voltage at the inverting input, so (with its high open-loop gain, A_{vol}) the output will remain saturated at $+V_{sat}$. For purposes of this discussion the amplifier is a type 741 operated at DC power supply potentials of ±12 Vdc, so V_{sat} will be about ±10 volts.

10-4.2 Transition state

The input trigger signal (V_t) is applied to the MMV of Fig. 10-4A through RC network $R4C2$. The general design rule for this network is that its time constant should be not more than one-tenth the time constant of the timing network:

$$R4C2 < \frac{R1C1}{10} \qquad (10\text{-}12)$$

At time $T1$ (see Fig. 10-4B) trigger signal V_t makes an abrupt HIGH to LOW transition to a peak value that is less than ($V1 - 0.7$) volts. Under this condition the polarity of V_{id} is now reversed and the inverting input now sees a positive voltage: ($V1 + V_t - 0.7$) is less than V_{D1}. The output voltage V_o now snaps rapidly to $-V_{sat}$. The fall time of the output signal is dependent upon the slew rate and the open-loop gain of the operational amplifier, $A1$.

10-4.3 Quasi-stable state

The output signal from the MMV is the quasi-stable state shown between $T1$ and $T2$ in Fig. 10-4B. It is called 'quasi-stable' because it does not change over $T = T2 - T1$, but when T expires the MMV 'times out' and V_o reverts to the stable state output voltage ($+V_{sat}$).

During the quasi-stable time $D1$ is reverse biased, and capacitor $C1$ discharges from $+0.7$ Vdc to zero and then recharges towards $-V_{sat}$. When $-V_o$ reaches $-V1$, however, the value of V_{id} crosses zero and that change forces V_o to snap once again to $+V_{sat}$.

Appealing to Eq. (10-6) makes it possible to derive the *timing equation* for the MMV. The timing capacitor must charge from an initial value (V_{C1})

to a final value (V_{C2}) in time T. The question is: 'What value of $R1C1$ will cause the required transitions?' Consider the case $R2 = R3$ ($V1 = 0.5\ V_{sat}$):

$$R1C1 = \frac{-T}{\ln\left[\dfrac{V_{sat} - V_{C2}}{V_{sat} - V_{C1}}\right]} \tag{10-13}$$

$$R1C1 = \frac{-T}{\ln\left[\dfrac{V_{sat} - ((0.5)(V_{sat} + 0.7))}{V_{sat} - 0.7}\right]} \tag{10-14}$$

and, for the case where $V_{sat} = 10$ Vdc:

$$R1C1 = \frac{-T}{\ln\left[\dfrac{10\ \text{Vdc} - ((0.5)(10\ \text{Vdc} + 0.7))}{10\ \text{Vdc} - 0.7}\right]} \tag{10-15}$$

$$R1C1 = \frac{-T}{\ln\left[\dfrac{10 - 5.35\ \text{volts}}{10\ \text{Vdc} - 0.7\ \text{volts}}\right]} \tag{10-16}$$

$$R1C1 = \frac{-T}{\ln\left[\dfrac{4.65}{9.3}\right]} \tag{10-17}$$

$$R1C1 = \frac{-T}{\ln(0.5)} \tag{10-18}$$

$$R1C1 = \frac{-T}{-0.69} \tag{10-19}$$

Thus,

$$T = 0.69R1C1 \tag{10-20}$$

EXAMPLE 10-2

A monostable multivibrator circuit is based on a 741 operational amplifier with an output saturation voltage of ±10 volts. Calculate the RC time constant needed to produce a 100 ms output pulse when feedback resistors $R2$ and $R3$ are each 10 kohms.

Solution

$$R1C1 = \frac{T}{0.69}$$

$$R1C1 = \frac{100\ \text{ms} \times \dfrac{1\ \text{s}}{1000\ \text{ms}}}{0.69}$$

$$R1C1 = \frac{0.1}{0.69} = 0.145\ \text{s} \qquad \blacksquare$$

Equation (10-20) represents the special case in which $\beta = 1/2$ (i.e. $R2 = R3$). Although $R2 = R3$ may be the usual case for this class of circuit, $R2$ and $R3$ might not be equal in other cases. A more generalized expression is:

$$RC = \frac{T}{\ln\left[\dfrac{1 + 0.7\,V/V_{sat}}{1 - \beta}\right]} \qquad (10\text{-}21)$$

in which,

$$\beta = \frac{R3}{R2 + R3} \qquad (10\text{-}22)$$

When the quasi-stable state times out, the circuit status returns to the stable state (where it remains dormant until triggered again).

EXAMPLE 10-3

A monostable multivibrator (Fig. 10-4A) is constructed with $R2 = 10$ kohms, and $R3 = 3.3$ kohms. The active device is a 741 operational amplifier operated such that $|V_{sat}| = 10$ volts. Calculate the RC time constant required to produce a 5 ms output pulse.

Solution

(A)   First calculate β:

$$\beta = \frac{R3}{R2 + R3}$$

$$\beta = \frac{3.3 \text{ kohms}}{10 \text{ kohms} + 3.3 \text{ kohms}}$$

$$\beta = \frac{3.3 \text{ kohms}}{13.3 \text{ kohms}} = 0.248$$

(B)   Calculate the RC time constant:

$$RC = \frac{T}{\ln\left[\dfrac{1 + 0.7\,V/V_{sat}}{1 - \beta}\right]}$$

$$RC = \frac{\left[5 \text{ ms} \times \dfrac{1 \text{ s}}{1000 \text{ ms}}\right]}{\ln\left[\dfrac{1 + 0.7\,V/10 \text{ volts}}{1 - 0.248}\right]}$$

$$RC = \frac{0.005 \text{ s}}{\ln[1.07/0.752]} = 0.0142 \text{ s} \qquad \blacksquare$$

10-4.4   Refractory period

At time $t2$ the output signal voltage V_o switches from $-V_{sat}$ to $+V_{sat}$. Although the output has timed out, the MMV is not yet ready to accept another trigger pulse. The *refractory state* between $t2$ and $t3$ is characterized by the output being in the stable state, but the input is unable to accept a new trigger input stimulus. The refractory period must await the discharge of $C1$ under the influence of the output voltage to satisfy $V1 < (V1 - 0.7)$ volts.

10-4.5   Retriggerable monostable multivibrators

The circuit of Fig. 10-4A is a nonretriggerable MMV. Once it is triggered the circuit will not respond to further trigger inputs until after both the quasi-stable and refractory states are completed. *A retriggerable monostable multivibrator* (RMMV) will respond to further trigger signals.

Figure 10-5 shows the RMMV response. An initial trigger signal (V_t) is received at time $t1$. The output snaps LOW and, under normal circumstances, it would remain in this quasi-stable state until time $t3$ when the duration T expires. But at time $t2$ a second trigger pulse is received. The circuit is now retriggered for another duration T, so will not time out until $t4$. The total time that the RMMV is in the quasi-stable state is $[T + (t2 - t1)]$. In other words, the RMMV output is active for the entire duration T plus that portion of the previous active time which expired when the next trigger pulse was received.

Figure 10-6A shows the circuit for a simple RMMV based on an operational amplifier. The two inputs are biased from a reference voltage source, $+V_{ref}$. The potential applied to $+IN$ is a fraction of $+V_{ref}$. That is $[(R3)(+V_{ref})/(R2 + R3)]$. The potential applied to $-IN$ is a function of $+V_{ref}$ and time constant $R1C1$. If the circuit is not triggered at turn-on,

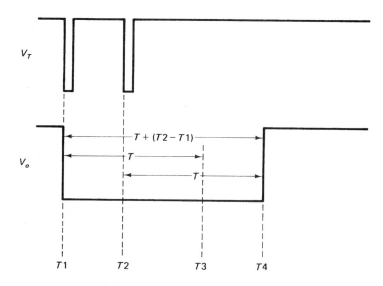

FIGURE 10-5   Trigger and output waveforms for retriggerable monostable multivibrator.

then the capacitor ($C1$) charges up to $+V_{ref}$, so $-$IN is more positive than $+$IN. This situation forces V_o to $-V_{sat}$, which is the stable state. When a positive going trigger pulse (V_t) is received (see Fig. 10-6B), it biases the junction field effect transistor (JFET), $Q1$, hard on. The JFET drain-source channel resistance drops to a very low value, causing $C1$ to discharge rapidly between $t1$ and $t2$. With V_c close to 0 Vdc, $+$IN is more positive than $-$IN, so the output snaps abruptly to $+V_{sat}$ at time $t1$. During the interval $t2$ to $t3$ capacitor $C1$ begins charging towards $+V_{ref}$, and V_o remains at $+V_{sat}$. Once V_c reaches $+V1$, however, the output of $A1$ snaps back to $-V_{sat}$.

The duration, T, is found from:

$$T = R1C1 \ln \left[\frac{R3}{R2} + 1 \right] \qquad (10\text{-}23)$$

EXAMPLE 10-4

Calculate the duration of a retriggerable monostable multivibrator (Fig. 10-6A) in which $R2 = R3 = 10$ kohms, $R1 = 15$ kohms and $C1 = 0.1$ μF.

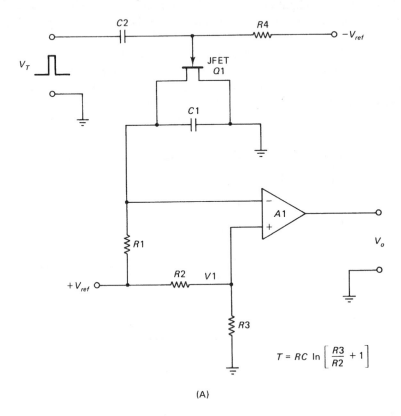

$$T = RC \ln \left[\frac{R3}{R2} + 1 \right]$$

(A)

FIGURE 10-6 Circuit and timing waveforms for retriggerable monostable multivibrator circuit.

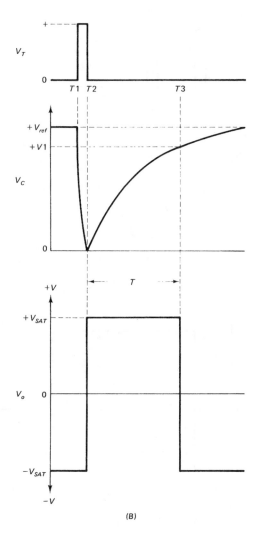

(B)

FIGURE 10-6   (*continued*)

Solution

$$T = R1C1 \ln\left[\frac{R3}{R2} + 1\right]$$

$$T = 15 \text{ k}\Omega \times \left[0.01 \text{ }\mu\text{F} \times \frac{1 \text{ F}}{10^6 \text{ }\mu\text{F}}\right] \times \ln\left[\frac{10 \text{ k}\Omega}{10 \text{ k}\Omega} + 1\right]$$

$$T = (1.5 \times 10^{-4}) \times \ln(2) \text{ s} = 1.04 \times 10^{-4} \text{ s} \qquad \blacksquare$$

The operation discussed above, and depicted in Fig. 10-6B, is for normal non-retriggered operation. Figure 10-6C shows the retriggered case. Here the

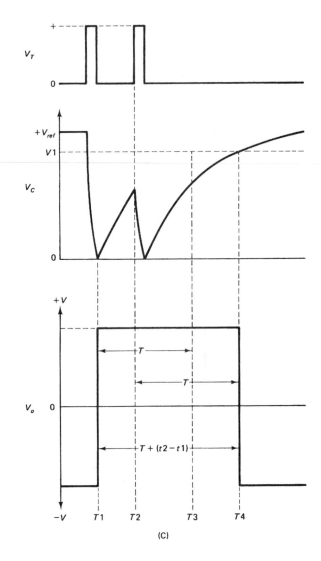

(C)

FIGURE 10-6 (continued)

RMMV receives a second trigger pulse at time $t2$, which forces the JFET ($Q1$) to turn on again, and rapidly discharge $C1$. The charging process then starts over again and will continue until the circuit times out... unless a further trigger pulse is received.

A common use for the RMMV is in alarm or sensing circuits. The RMMV is triggered by some external event, and will continually retrigger as long as the external event keeps occurring. But if no event is sensed prior to time-out, the RMMV returns to the stable state, and the following circuitry will be triggered to alarm status. An example is a medical respirator alarm. A sensor in the respirator line senses variations in either pressure or air temperature caused by breathing. Each time a breath is sensed it retriggers the RMMV. But if the patient ceases breathing, and the respirator fails to augment breathing, then the RMMV will time out and cause an alarm to nearby medical personnel.

10-5 ASTABLE (FREE-RUNNING) CIRCUITS ⎯⎯⎯⎯⎯⎯⎯⎯

The circuits discussed in the previous section were aperiodic, i.e. an output pulse occurs only once in response to a stimulus or trigger. Such circuits are said to be *monostable* because they possess only one stable state. An *astable multivibrator* (AMV), on the other hand, is free-running. The output of the AMV is a pulse or wave train that is *periodic*. In a periodic signal the wave repeats itself indefinitely until the circuit is either turned off or otherwise inhibited.

Astable multivibrators are oscillators. Waveforms available from the AMV include squarewaves, triangle waves, and sawtooth waves. Sinewaves are also available from oscillator circuits, but those circuits operate differently from the others so are handled separately.

10-5.1 Non-sinusoidal waveform generators

The non-sinusoidal AMV circuits will produce square, triangle or sawtooth waves. When combined with a monostable multivibrator (MMV), a pulse generator results. Because the squarewave generator is the most basic form, the discussion of AMV circuits begins with squarewaves.

10-5.2 Squarewave generators

Figure 10-7 shows the basic squarewave. Each time interval of the wave is quasi-stable, so one may conclude that the squarewave generator has no stable states (hence it is called *astable*). The waveform snaps back and forth between $-V$ and $+V$, dwelling on each level of a duration of time (t_a or t_b). The period, T, is:

$$T = t_a + t_b \qquad (10\text{-}24)$$

where:
    T is the period of the squarewave ($t1$ to $t3$)
    t_a is the interval $t1$ to $t2$
    t_b is the interval $t2$ to $t3$

The frequency of oscillation (F) is the reciprocal of T:

$$F = \frac{1}{T} \qquad (10\text{-}25)$$

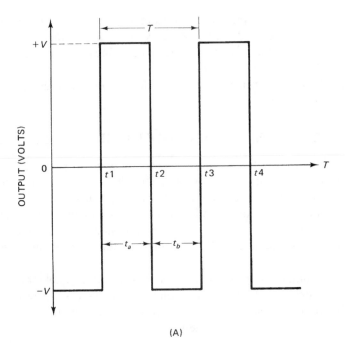

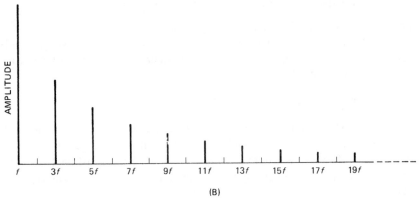

FIGURE 10-7   (A) Squarewave; (B) frequency spectrum for squarewave.

The ideal squarewave is both base-line and time-line symmetrical. That means that $|+V| = |-V|$ and $t_a = t_b$. Under time-line symmetry $t_a = t_b = t$, so $T = 2t$ and $f = 1/2t$.

All continuous mathematical functions can be constructed from a fundamental frequency sinewave (f) added to a series of sine and cosine harmonics ($2f, 3f, 4f, \ldots, nf$). The mathematical expression of which harmonics are present, their respective amplitudes and phase relationships, is called the

Fourier series of the waveform. The Fourier series is usually depicted as a bar graph spectrum such as Fig. 10-7B.

In the ideal, symmetrical, squarewave the Fourier spectrum (Fig. 10-7B) consists of the fundamental frequency (f) plus the *odd-order* harmonics ($3f$, $5f$, $7f$, etc.). Furthermore, the harmonics are in-phase with the fundamental. Theoretically, an infinite number of odd-number harmonics are present in the ideal squarewave. However, in practical squarewaves the 'ideal' is considered satisfied with harmonics to about $1000f$. That ideal is almost never reached, however, due to the normal bandwidth limitations of the circuit. An indicator of harmonic content is the risetime of the squarewaves: the faster the risetime, the higher the number of harmonics.

The circuit for an operational amplifier squarewave generator is shown in Fig. 10-8A. The basic circuit is similar to the simple voltage comparator and the MMV (see Section 10-4). Like the MMV, the AMV operation depends upon the relationship between V_{-IN} and V_{+IN}. In the circuit of Fig. 10-8A the voltage applied to the noninverting input (V_{+IN}) is determined by a resistor voltage divider, $R2$ and $R3$. This voltage is called $V1$ in Fig. 10-8A and is:

$$V1 = \frac{V_o R3}{R2 + R3} \qquad (10\text{-}26)$$

or, when V_o is saturated,

$$V1 = \frac{V_{sat} R3}{R2 + R3} \qquad (10\text{-}27)$$

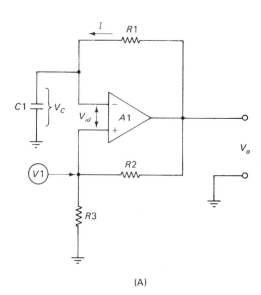

(A)

FIGURE 10-8 (A) Astable multivibrator circuit; (B) timing waveforms.

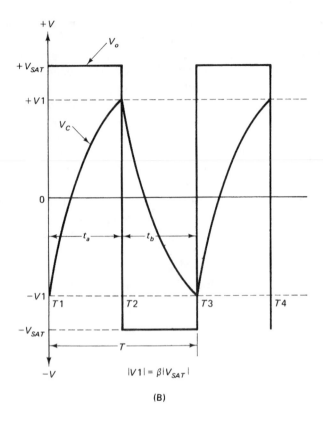

(B)

FIGURE 10-8 *(continued)*

Once again, the factor $R3/(R2 + R3)$ is often designated β:

$$\beta = \frac{R3}{R2 + R3} \qquad (10\text{-}28)$$

Because Eq. (10-20) is always a fraction $V1 < V_{sat}$ and $V1$ is of the same polarity as V_{sat}.

The voltage applied to the inverting input (V_{-IN}) is the voltage across capacitor $C1$, i.e. V_{C1}. This voltage is created when $C1$ charges under the influence of current I, which in turn is a function of V_o and the time constant $R1C1$. Timing operation of the circuit is shown in Fig. 10-8B.

At turn-on, $V_{C1} = 0$ volts and $V_o = +V_{sat}$, so $V1 = +V1 = \beta(+V_{sat})$. Because $V_{C1} < V1$, the op-amp sees a negative differential input voltage so the output remains at $+V_{sat}$. During this time, however, V_{C1} is charging towards $+V_{sat}$ at a rate of:

$$V_{C1} = V_{sat}(1 - e^{-t_a/R1C1}) \qquad (10\text{-}29)$$

When V_{C1} reaches $+V1$, however, the op-amp sees $V_{C1} = V1$, so $V_{id} = 0$. The output now snaps from $+V_{sat}$ to $-V_{sat}$ (time $t2$ in Fig. 10-8B). The capacitor now begins to discharge from $+V1$ towards zero, and then recharges towards $-V_{sat}$. When it reaches $-V1$, the inputs are once again zero so the output again snaps to $+V_{sat}$. The output continuously snaps back and forth between $-V_{sat}$ and $+V_{sat}$, thereby producing a squarewave output signal.

Again appealing to Eq. (10-6) to find the time constant required to charge from an initial voltage V_{C1} to an end voltage V_{C2} in time t:

$$RC = \frac{-T}{\ln\left[\dfrac{V - V_{C2}}{V - V_{C1}}\right]} \tag{10-30}$$

In Fig. 10-8A the RC time constant is $R1C1$. From Fig. 10-8B it is apparent that, for interval t_a, $V_{C1} = -\beta V_{sat}$, $V_{C2} = +\beta V_{sat}$ and $V = V_{sat}$. To calculate the period T, then:

$$2R1C1 = \frac{-T}{\ln\left[\dfrac{V_{sat} - \beta V_{sat}}{V_{sat} - (-\beta V_{sat})}\right]} \tag{10-31}$$

or, rearranging Eq. (10-31):

$$-T = 2R1C1\ln\left[\frac{V_{sat} - \beta V_{sat}}{V_{sat} - (-\beta V_{sat})}\right] \tag{10-32}$$

$$-T = 2R1C1 = 2R1C1\ln\left[\frac{1 - \beta}{1 + \beta}\right] \tag{10-33}$$

$$T = 2R1C1 = 2R1C1\ln\left[\frac{1 + \beta}{1 - \beta}\right] \tag{10-34}$$

Because $\beta = R3/(R2 + R3)$:

$$T = 2R1C1\ln\left[\frac{1 + (R3/(R2 + R3))}{1 - (R3/(R2 + R3))}\right] \tag{10-35}$$

which reduces to:

$$T = 2R1C1\ln\left[\frac{2R2}{R3}\right] \tag{10-36}$$

EXAMPLE 10-5

Calculate the oscillating frequency for an astable multivibrator in which $R1 = 10$ kohms, $R2 = 10$ kohms, and $C1 = 0.005$ μF.

Solution

$$T = 2R1C1\ln\left[\frac{2R2}{R3}\right]$$

$$T = (2)(10^4 \ \Omega) \left[0.005 \ \mu F \times \frac{1 \ F}{10^6 \ \mu F} \right] \ln \left[\frac{(2)(10 \ k\Omega)}{4.7 \ k\Omega} \right]$$

$$T = (0.0001) \ln(4.26) \ s = 0.000\,145 \ s \qquad\blacksquare$$

Equation (10-36) defines the frequency of oscillation for any combination of $R1$, $R2$, $R3$ and $C1$. In the special case $R2 = R3$, $\beta = 0.5$, so:

$$T = 2R1C1 \ln \left[\frac{1 + 0.5}{1 - 0.5} \right] \qquad (10\text{-}37)$$

$$T = 2R1C1 \ln \left[\frac{1.5}{0.5} \right] \qquad (10\text{-}38)$$

$$T = 2R1C1 \ln(3) = 2.2 \ R1C1 \qquad (10\text{-}39)$$

EXAMPLE 10-6

A squarewave oscillator is constructed such that $R2 = R3$. Calculate the time constant required for a 1000 Hz symmetrical squarewave.

Solution

$$R1C1 = \frac{1}{2.2f}$$

$$R1C1 = \frac{1}{(2.2)(1000 \ Hz)} = 0.000\,46 \ s \qquad\blacksquare$$

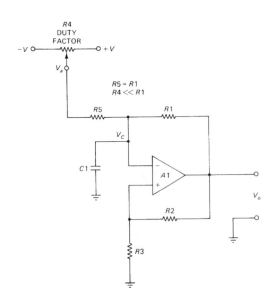

FIGURE 10-9   Variable duty cycle astable multivibrator.

The circuit of Fig. 10-8A produces time-line symmetrical squarewaves (i.e. $t_a = t_b$). If time-line asymmetrical squarewaves are required, then a circuit such as either Fig. 10-9 or 10-11A is required. The circuit in Fig. 10-9 uses a potentiometer ($R4$) and a fixed resistor ($R5$) to establish a variable duty cycle asymmetry. The circuit is similar to Fig. 10-8A, but with an offset circuit ($R4/R5$) added. The assumptions are $R5 = R1$, and $R4 \ll R1$. If V_a is the potentiometer output voltage, $C1$ charges at a rate of $(R1/2)C1$ towards a potential of $(V_a + V_{sat})$. After output transition, however, the capacitor discharges at the same $(R1/2)C1$ rate towards $(V_a - V_{sat})$. The two interval times are therefore different; t_a and t_b are no longer equal. Figure 10-10 shows three extremes of V_a: $V_a = +V$ (Fig. 10-10A), $V_a = 0$ (Fig. 10-10B) and $V_a = -V$ (Fig. 10-10C). These traces represent very long, equal and very short duty cycles, respectively.

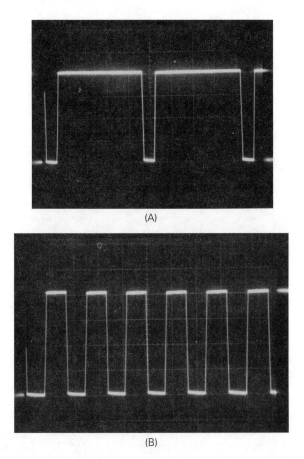

(A)

(B)

FIGURE 10-10 Output waveforms of variable duty cycle astable multivibrator at various settings of potentiometer.

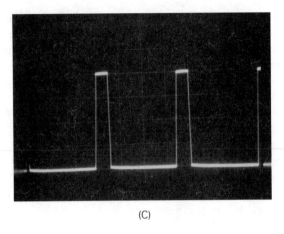

(C)

FIGURE 10-10 (*continued*)

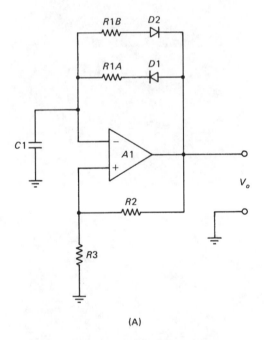

(A)

FIGURE 10-11 (A) Fixed duty cycle circuit; (B) waveform when *R*1A/B ratio is 3:1; (C) when *R*1A/B ratio is 10:1.

The circuit of Fig. 10-11A also produces asymmetrical squarewaves, but the duty cycle is fixed instead of variable. Once again the basic circuit is like Fig. 10-8A, but with added components. In Fig. 10-11A the *RC* timing network is altered such that the resistors are different on each swing of the output signal. During t_a, $V_a = +V_{sat}$ so diode *D*1 is forward biased and *D*2

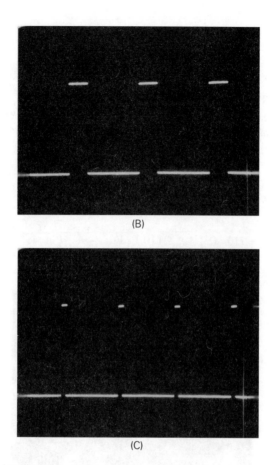

(B)

(C)

FIGURE 10-11 (continued)

is reverse biased. For this interval:

$$t_a = (R1A)(C1) \ln \left[1 + \frac{2R2}{R3} \right] \qquad (10\text{-}40)$$

During the alternate halfcycle (t_b), the output voltage V_o is at $-V_{sat}$, so $D1$ is reverse biased and $D2$ is forward biased. During this interval $R1B$ is the timing resistor, while $R1A$ is effectively out of the circuit. The timing equation is:

$$t_a = (R1B)(C1) \ln \left[1 + \frac{2R2}{R3} \right] \qquad (10\text{-}41)$$

The total period, T, is $t_a + t_b$, so:

$$t_a = (R1A)(C1) \ln \left[1 + \frac{2R2}{R3} \right] + (R1B)(C1) \ln \left[1 + \frac{2R2}{R3} \right] \qquad (10\text{-}42)$$

Collecting terms:

$$T = (R1A + R1B)(C1) \ln \left[1 + \frac{2R2}{R3} \right] \qquad (10\text{-}43)$$

EXAMPLE 10-7

An asymmetrical astable multivibrator is designed such that $R1A = 100$ kohms, $R1B = 10$ kohms, $C1 = 0.1$ μF, and $R2 = R3$. Calculate the period of oscillation.

Solution

Let $R2 = R3 = R$

$$T = (R1A + R1B)(C1) \ln \left[1 + \frac{2R2}{R3} \right]$$

$$T = (10^5 \ \Omega + 10^4 \ \Omega) \left[0.1 \ \mu F \times \frac{1 \ F}{10^6 \ \mu F} \right] \ln \left[1 + \frac{2R}{R} \right]$$

$$T = 0.011 \ln(3) = 0.012 \text{ s} \qquad \blacksquare$$

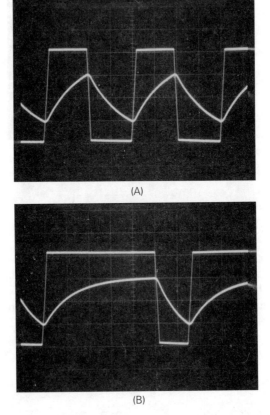

(A)

(B)

FIGURE 10-12   Output voltage and capacitor voltage waveforms for different $R1A/B$ ratios.

Equation (10-43) defines the oscillation frequency of the circuit in Fig. 10-11A. Figures 10-11B and 10-11C show the effects of two values of $R1A/R1B$ ratio. In Fig. 10-11B the ratio $R1A/R1B = 3{:}1$, while in Fig. 10-11C $R1A/R1B = 10{:}1$.

The effect of this circuit on capacitor charging can be seen in Fig. 10-12. A relatively low $R1A/R1B$ ratio is seen in Fig. 10-12A. Notice in the lower trace that the capacitor charge time is long compared with the discharge time. The effect is seen in even better for the case of a high $R1A/R1B$ ratio (Fig. 10-12B).

Output voltage limiting. The standard op-amp MMV or AMV circuit sometimes produces a relatively sloppy square output wave. By adding a pair of back-to-back zener diodes (Fig. 10-13A) across the output, however, the signal can be cleaned up. For each polarity the output signal sees one forward biased and one reverse biased zener diode. On the positive swing, the output voltage is clamped at $[V_{Z1} + 0.7]$ volts. The 0.7 volts factor represents the normal junction potential across the forward biased diode ($D2$). On negative swings of the output signal, the situation reverses. The output signal is clamped to $[-(V_{Z2} + 0.7)]$ volts. Figure 10-13B shows the unclamped output signal of a squarewave generator. The signal swings ± 10 volts. Figure 10-13C shows the output of the same circuit when a pair of 5.6 volt zener diodes are connected across the output. The output is reduced to $\pm[5.6 + 0.7]$ volts, but the corners are sharper.

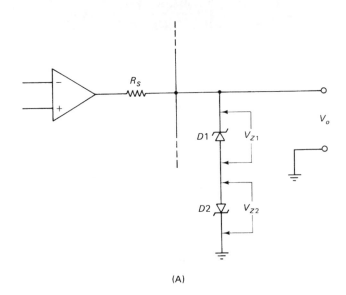

(A)

FIGURE 10-13   (A) Output voltage limiting circuit; (B) output waveform without diodes; (C) output waveform with diodes.

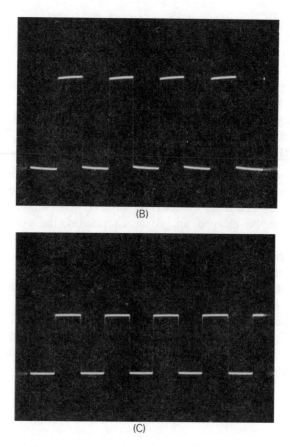

(B)

(C)

FIGURE 10-13   (continued)

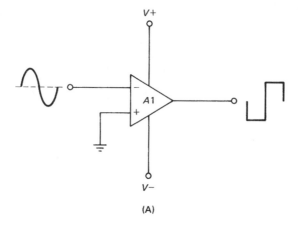

V+

A1

V−

(A)

FIGURE 10-14   (A) circuit to generate squarewaves from sinewave input; (B) input and output waveforms.

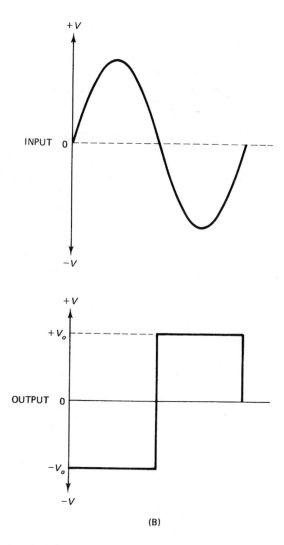

INPUT

OUTPUT

(B)

FIGURE 10-14 *(continued)*

Squarewaves from sinewaves. Figure 10-14 shows a method for converting sinewaves to squarewaves. The circuit is shown in Fig. 10-14A, while the waveforms are shown in Fig. 10-14B. The circuit is an operational amplifier connected as a comparator. Because the op-amp has no negative feedback path, the gain is extremely high (i.e. A_{vol}); in typical op-amps gains of 20 000 to 2 000 000 are found. Thus, a voltage difference across the input terminals of only a few millivolts will saturate the output. From this behavior, the operation of the circuit, and the waveform in Fig. 10-14B can be understood.

The input waveform is a sinewave. Because the noninverting input is grounded (Fig. 10-14A), the output of the op-amp is zero only when the input signal voltage is also zero. When the sinewave is positive, the output signal will be at $-V_o$; when the sinewave is negative, the output signal will be at $+V_o$. The output signal will be a squarewave at the sinewave frequency, with a peak-to-peak amplitude of $[(+V_o) - (-V_o)]$.

10-5.3   Triangle and sawtooth waveform generators

Triangle and *sawtooth waveforms* (Fig. 10-15) are examples of periodic ramp waveforms. The sawtooth (Fig. 10-15A) is a single ramp waveform. The voltage begins to rise linearly at time $t1$. At time $t2$ the waveform abruptly drops back to zero, where it again starts to ramp up linearly. The sawtooth is periodic, and the period is defined as T (see Fig. 10-15A), so the frequency is $1/T$.

The triangle waveform (Fig. 10-15B) is a double ramp. The waveform begins to ramp up linearly at time $t1$. It reverses direction at time $t2$, and then ramps downward linearly until time $t3$. At time $t3$ the waveform again

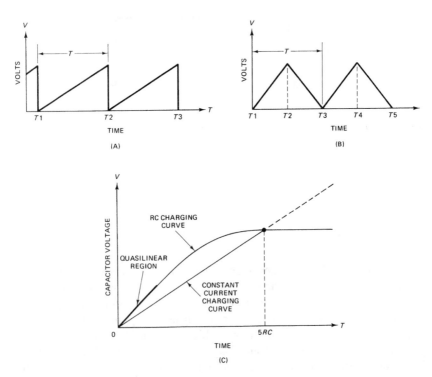

FIGURE 10-15   (A) Sawtooth waveform; (B) triangle waveform; (C) *RC* charging with normal and constant drive.

reverses to begin ramping upwards. The period of the triangle waveform (T) is $T1 - T3$.

Ramp generator circuits are derived from capacitor charging mechanisms. The familiar RC charging curve was discussed earlier in this chapter, and is reproduced in Fig. 10-15C. The RC charging waveform has an exponential shape, so is not well suited to generating the ramp function. There are two approaches to forcing the capacitor charging waveform to be more linear. The first is to limit the charging time to the short quasi-linear segment shown in Fig. 10-15C. The ramp thus obtained is not very linear, is limited in amplitude to a small fraction of $V1$, and has a relatively steep slope. A superior method is to charge the capacitor through a constant current source (CCS). Using the CCS results in the linear ramp shown in Fig. 10-15C.

Triangle and sawtooth waveform generator circuits implement the constant current ramp by using a Miller integrator circuit to charge the capacitor (Fig. 10-16A). When a Miller integrator is driven by a stable DC reference voltage, the output is a linearly rising ramp. The ramp voltage (V_o) is:

$$V_o = \frac{V_{ref}}{T} \qquad (10\text{-}44)$$

or, because $T = RC$:

$$V_o = \frac{(V_{ref})}{RC} \qquad (10\text{-}45)$$

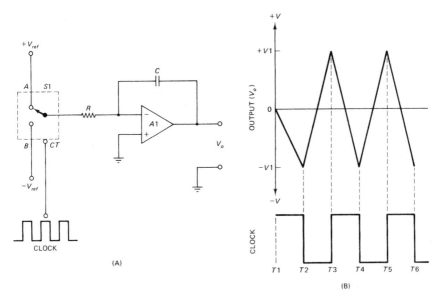

FIGURE 10-16   Switching approach to triangle waveform generation: (A) circuit; (B) timing waveform.

If $V_{ref} = +10$ Vdc, and the RC time constant is $T = RC = 1$ second, the ramp slope is:

$$V_o = \frac{10 \text{ volts}}{1 \text{ s}} \qquad (10\text{-}46)$$

$$V_o = 10 \text{ volts/s} \qquad (10\text{-}47)$$

Triangle generators. Figure 10-16A shows a simplified circuit model of a triangle waveform generator. This circuit consists of a Miller integrator as the ramp generator, and an SPDT switch ($S1$) that can select either positive ($+V_{ref}$) or negative ($-V_{ref}$) reference voltage sources.

For purposes of discussion, switch $S1$ is an electronic switch that is toggled back and forth between positions A and B by a squarewave applied to the control terminal (CT). Assume an initial condition at time $t2$, at which point $V_o = -V1$ and the input of the integrator is connected to $-V_{ref}$. At time $t2$ the squarewave switch driver changes to the opposite state, so $S1$ toggles to connect $+V_{ref}$ to the integrator input. The ramp output will rise linearly at a rate of $+V_{ref}/RC$ until the switch again toggles at time $t3$. At this point, the ramp is under the influence of $-V_{ref}$, so drops linearly from $+V1$ to $-V1$. The switch continuously toggles back and forth between $-V_{ref}$ and $+V_{ref}$, so the output voltage (V_o) continuously ramps back and forth between $-V1$ and $+V1$.

The circuit of Fig. 10-16A is not practical, but serves as an analogy for the actual circuit. Figure 10-17A shows the circuit for a triangle waveform generator in which a Miller integrator forms the ramp generator and a voltage comparator serves as the switch. The comparator uses the positive feedback configuration, so operates as a noninverting Schmitt trigger. Such a circuit snaps HIGH ($V_B = +V_{sat}$) when the input signal crosses a certain threshold voltage in the positive going direction. It will snap LOW ($V_B = -V_{sat}$) when the input signal crosses a second threshold in a negative going direction. The two thresholds may or may not be the same potential.

Because zener diodes $D1$ and $D2$ are in the circuit, the maximum allowable value of $+V_B$ is $[V_{ZD1} + 0.7]$ volts, while the limit for $-V_B$ is $-[V_{ZD2} + 0.7]$ volts. Assume that $V_{ZD1} = V_{ZD2}$, so $|+V_B| = |-V_B|$. These potentials represent $\pm V_{ref}$ discussed in the analogy presented above, so are the potentials that affect the ramp generator input.

Consider an initial state in which V_B is at the negative limit $-V_B$. The output V_o will begin to ramp upwards from a minimum voltage of:

$$V1 = \frac{V_A(R2 + R4)}{R4} - \frac{V_B R2}{R4} \qquad (10\text{-}48)$$

The output will continue to ramp upwards towards a maximum value of:

$$V3 = \frac{V_A(R2 + R4)}{R4} + \frac{V_B R2}{R4} \qquad (10\text{-}49)$$

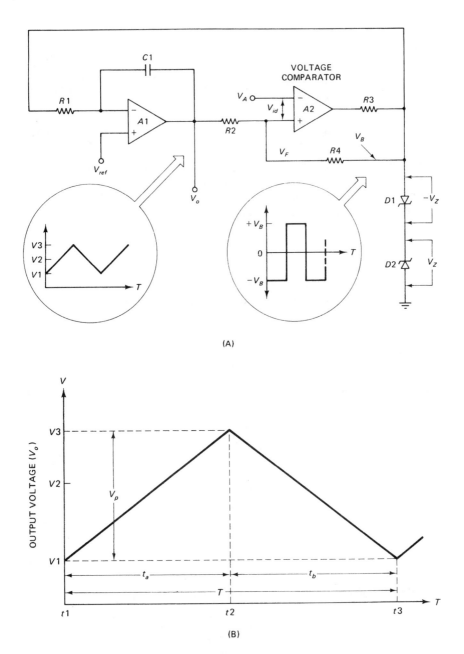

FIGURE 10-17 (A) Triangle waveform generator circuit; (B) output waveform.

causing a peak swing voltage of:

$$V_p = V3 - V1 \tag{10-50}$$

$$V_p = \left[\frac{V_A(R2 + R4)}{R4} + \frac{V_B R2}{R4} \right] - \left[\frac{V_A(R2 + R4)}{R4} - \frac{V_B R2}{R4} \right] \tag{10-51}$$

$$V_p = \frac{V_B R2}{R4} + \frac{V_B R2}{R4} = \frac{2V_B}{R4} \tag{10-52}$$

Switching of the comparator occurs when the differential input voltage V_{id} is zero. The inverting input $(-IN)$ voltage is V_A, which is a fixed reference potential. The noninverting input $(+IN)$ is at a voltage (V_F) that is the superposition of two voltages, V_o and V_B:

$$V_F = \frac{V_o R4}{R2 + R4} \pm \frac{\pm V_B R2}{R2 + R3} \tag{10-53}$$

If $+V_B = -V_B$, then the positive and negative thresholds are equal. The duration of each ramp (t_a and t_b) can be found from:

$$t_{a,b} = \frac{V_p}{\left[\dfrac{V_B}{R1C1} \right]} \tag{10-54}$$

The value of V_B is selected from $-V_B$ or $+V_B$ as needed. In Eq. (10-52) it was found that $V_p = 2V_B R2/R4$, so:

$$t_{a,b} = \frac{\left[\dfrac{2V_B R2}{R4} \right]}{\left[\dfrac{V_B}{R1C1} \right]} \tag{10-55}$$

$$t_{a,b} = \frac{R1C1}{V_B} \times \frac{2V_B R2}{R4} \tag{10-56}$$

$$t_{a,b} = R1C1 \left[\frac{2R2}{R4} \right] \tag{10-57}$$

or, in the less general (but more common) case of $t_a = t_b$:

$$T = 2R1C1 \left[\frac{2R2}{R4} \right] \tag{10-58}$$

The frequency of the triangle wave is the reciprocal of the period $(1/T)$, so:

$$F = \frac{1}{T} \tag{10-59}$$

$$F = \cfrac{1}{\left[\cfrac{4R1C1R2}{R4}\right]} \qquad (10\text{-}60)$$

$$F = \cfrac{R4}{4R1C1R2} \qquad (10\text{-}61)$$

Sawtooth generators. The sawtooth wave (seen previously in Fig. 10-15A) is a single slope ramp function. The wave ramps linearly upwards or downwards, and then abruptly snaps back to the initial baseline condition. Figure 10-18A shows a simple sawtooth waveform generator circuit. A constant current source charges a capacitor which generates the linear ramp function (Fig. 10-18B). When the ramp voltage (V_c) reaches the maximum point (V_p) switch $S1$ is closed, forcing V_c back to zero by discharging the capacitor. If switch $S1$ remains closed, the sawtooth is terminated. But if $S1$ reopens, however, a second sawtooth is created as the capacitor recharges.

Figure 10-19A shows the circuit for a periodic sawtooth oscillator. It is similar to Fig. 10-18A except a junction field effect transistor (JFET), $Q1$, is used as the discharge switch. When $Q1$ is turned off, the output voltage ramps upwards (see Fig. 10-19B). When the gate is pulsed hard-on, the drain-source channel resistance drops from a very high value to a very low value, forcing $C1$ to rapidly discharge. In the absence of a gate pulse, however, the channel resistance remains very high. At time $t1$ the gate is turned off, so V_c begins ramping upwards. At $t2$ the JFET gate is pulsed, so $C1$ rapidly discharges back to zero. When the pulse ($t2 - t3$) ends, however, $Q1$ turns off again and the ramp starts over. The same circuit can be used for single sweep operation by replacing the pulse train applied to the gate of $Q1$ with the output of a monostable multivibrator.

The circuit of Fig. 10-20A shows a sawtooth generator which uses a Miller integrator ($A1$) as a ramp generator and replaces the discharge switch with an

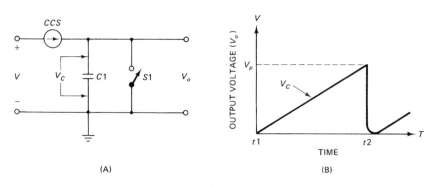

FIGURE 10-18   (A) Switching approach to sawtooth generation; (B) output waveform.

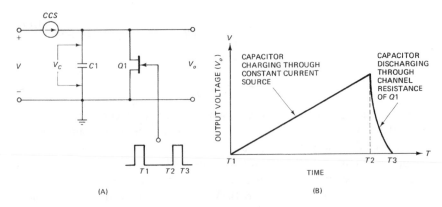

FIGURE 10-19 (A) CMOS switch approach to sawtooth generation; (B) output waveforms.

electronic switch driven by a voltage comparator and one-shot circuit. The timing diagram for this circuit is shown in Fig. 10-20B. Under the initial condition at time $t1$, the output voltage (V_o) ramps upwards at a rate of $[-(-V_{ref})/R1C1]$. The voltage comparator $(A2)$ is biased with the noninverting input $(+IN)$ set to $V1$ and the inverting input $(-IN)$ set to V_o. The comparator differential input voltage $V_{id} = (V1 - V_o)$. As long as $V1 > V_o$ the comparator sees a negative input and produces a HIGH output of $+V_{sat}$. At the point where $V1 = V_o$, the differential input voltage is zero, so the

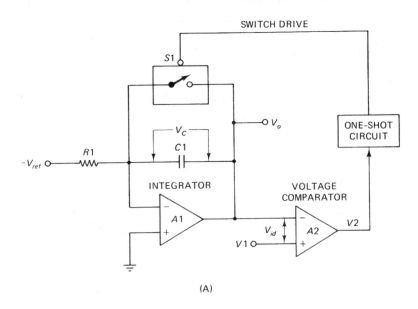

FIGURE 10-20 (A) Integrator–comparator approach to sawtooth generation; (B) timing waveforms.

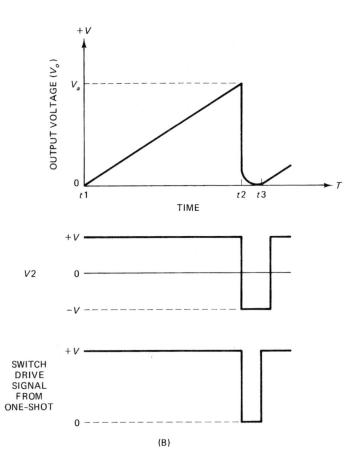

(B)

FIGURE 10-20 (continued)

output of $A2$ (i.e. voltage $V2$) drops LOW ($-V_{sat}$). The negative going edge of $V2$ at time $t2$ triggers the one-shot circuit. The output of the one-shot briefly closes electronic switch $S1$, causing the capacitor to discharge. The one-shot pulse ends at time $t3$, so $S1$ reopens and allows V_o to ramp upwards again.

The *staircase generator* (Fig. 10-21A) is a variant of the sawtooth generator circuit. The input amplifier ($A1$) provides buffering. A squarewave clock signal applied to the input of $A1$ is passed through capacitor $C2$ to a diode clipping network ($D1$, $D2$). The clipping circuit removes the negative excursions of the squarewave (see inset to Fig. 10-21A). The remaining positive polarity pulses are applied to the input of the inverting Miller integrator ramp generator circuit. Each pulse adds a slight step increase to the capacitor charge voltage, and unless there is significant droop between pulses, the output will ramp up to a negative potential in the staircase fashion shown in the inset to Fig. 10-21B.

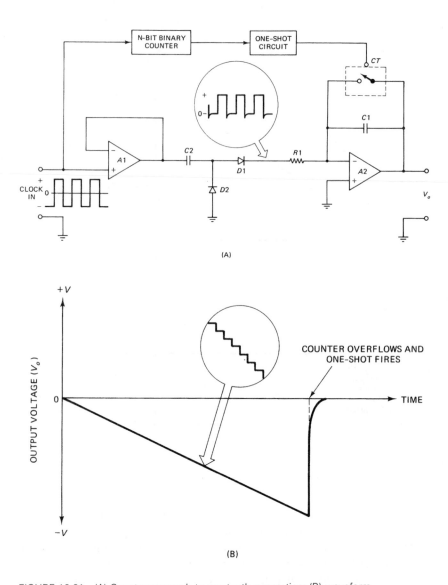

FIGURE 10-21 (A) Counter approach to sawtooth generation; (B) waveform.

The reset circuitry here is a little different. Although the comparator method of Fig. 10-20A would also work, this circuit takes advantage of the input squarewave to provide the period timing of the sawtooth. The squarewaves are applied to the input of an N-bit binary digital counter circuit. When 2^N pulses have passed, the counter overflows on $2^N + 1$ and triggers a one-shot circuit. As in the previous case, the one-shot output pulse momentarily closes the electronic reset switch shunted across capacitor $C1$.

10-6 FEEDBACK OSCILLATORS

A feedback oscillator (Fig. 10-22) consists of an amplifier with an open-loop gain of A_{vol} and a feedback network with a gain or transfer function β. It is called a 'feedback oscillator' because the output signal of the amplifier is fed back to the amplifier's own input by way of the feedback network. Figure 10-22 is a block diagram model of the feedback oscillator. That it bears more than a superficial resemblance to a feedback amplifier is no coincidence. Indeed, as anyone who has misdesigned or misconstructed an amplifier knows all too well, a feedback oscillator is an amplifier in which special conditions prevail. These conditions are called *Barkhausen's criteria for oscillation*:

1. Feedback voltage V_F must be in-phase (360°) with the input voltage, and
2. The loop gain βA_{vol} must be unity (1) or greater.

The first of these criteria means that the total phase shift from the input of the amplifier, to the output of the amplifier, around the loop back to the input, must be 360° (2π radians) or an integer (N) multiple of 360° (i.e. $N2\pi$ radians).

The amplifier can be any of many different devices. In some circuits it will be a common emitter bipolar transistor (NPN or PNP devices). In others it will be a junction field effect transistor (JFET) or metal oxide semiconductor field effect transistor (MOSFET). In older equipment it was a vacuum tube. In modern circuits the active device will probably be either an integrated circuit operational amplifier, or some other form of linear IC amplifier.

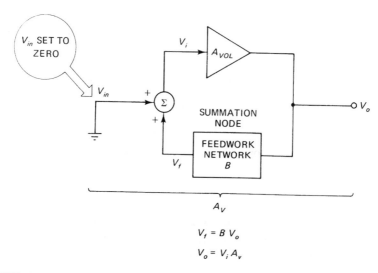

FIGURE 10-22   Feedback oscillator block diagram.

The amplifier is most frequently an inverting type, so the output is out of phase with the input by 180°. As a result, in order to obtain the required 360° phase shift, an additional phase shift of 180° must be provided in the feedback network *at the frequency of oscillation only*. If the network is designed to produce this phase shift at only one frequency, then the oscillator will produce a sinewave output on that frequency.

Before considering specific sinewave oscillator circuits let's examine Fig. 10-22 more closely. Several things can be determined about the circuit:

$$V_i = V_{in} + V_F \qquad (10\text{-}62)$$

so,

$$V_{in} = V_i + V_F \qquad (10\text{-}63)$$

and also,

$$V_F = \beta V_o \qquad (10\text{-}64)$$

$$V_o = V_i A_{vol} \qquad (10\text{-}65)$$

The transfer function (or gain) A_v is:

$$A_v = \frac{V_o}{V_{in}} \qquad (10\text{-}66)$$

Substituting Eqs (10-63) and (10-65) into Eq. (10-66):

$$A_v = \frac{V_i A_{vol}}{V_i - V_F} \qquad (10\text{-}67)$$

From Eq. (10-64), $V_F = \beta V_o$, so

$$A_v = \frac{V_i A_{vol}}{V_i - \beta V_o} \qquad (10\text{-}68)$$

But Eq. (10-65) shows $V_o = V_i A_{vol}$, so Eq. (10-68) can be written:

$$A_v = \frac{V_i A_{vol}}{V_i - \beta V_i A_{vol}} \qquad (10\text{-}69)$$

and, dividing both numerator and denominator by V_i:

$$A_v = \frac{A_{vol}}{1 - \beta A_{vol}} \qquad (10\text{-}70)$$

Equation (10-70) serves for both feedback amplifiers and oscillators. But in the special case of an oscillator $V_{in} = 0$, so $V_o \to \infty$. Implied, therefore, is that the denominator of Eq. (10-70) must also be zero:

$$1 - \beta A_{vol} = 0 \qquad (10\text{-}71)$$

Therefore, for the case of the feedback oscillator:

$$\beta A_{vol} = 1 \qquad (10\text{-}72)$$

βA_{vol} is the loop gain of the amplifier and feedback network, so Eq. (10-72) meets Barkhausen's second criterion.

10-7 SINEWAVE OSCILLATORS

Sinewave oscillators produce an output signal that is sinusoidal. Such a signal is ideally very pure, and if indeed it is perfect, then its Fourier spectrum will contain only the fundamental frequency and no harmonics. It is the harmonics in a non-sinusoidal waveform that give it a characteristic shape. The active element in the circuits described in this circuit is the operational amplifier. However, any linear amplifier will also work in place of the operational amplifier. The one circuit that shows the principles most clearly is the *RC phase shift oscillator*, so it is with that circuit that the discussion starts.

Stability in oscillator circuits can refer to several different phenomena. First, is *frequency stability*, which refers to the ability of the oscillator to remain on the design frequency over time. Several different factors affect frequency stability, but the most important are *temperature* and *power supply voltage variations*.

Another form of stability is *amplitude stability*. Because sinewave oscillators do not operate in the saturated mode, it is possible for minor variations in circuit gain to affect the amplitude of the output signal. Again the factors most often cited for this problem include temperature and DC power supply variations. The latter is overcome by using regulated DC power supplies for the oscillator. The former is overcome by either temperature compensated design or maintaining a constant operating temperature. Some variable sinewave oscillators will exhibit amplitude variation of the output signal when the operating frequency is changed. In these circuits either a self-compensation element is used, or an *automatic level control* amplifier stage is used.

Still another form of stability regards the purity of the output signal. If the circuit exhibits spurious oscillations, then these will be superimposed on the output signal. As with any circuit containing an op-amp, or any other high gain linear amplifier, it is necessary to properly decouple the DC power supply lines. It may also be necessary to frequency compensate the circuit.

10-7.1 RC phase shift oscillator circuits

The *RC* phase shift oscillator is based on a three-stage cascade resistor–capacitor network such as shown in Fig. 10-23A. An *RC* network will exhibit a phase shift ϕ (Fig. 10-23B) that is a function of resistance (R) and capacitive reactance (X_c). Because X_c is inversely proportional to frequency ($1/2\pi f C$), the phase angle is therefore a function of frequency. The goal in designing the *RC* phase shift oscillator is to create a phase

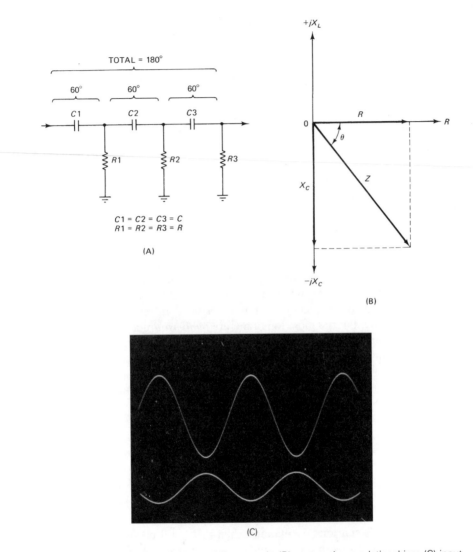

FIGURE 10-23 (A) *RC* 180° phase shift network; (B) vector phase relationships; (C) input and output circuits.

shift of 180° between the input and output of the network at the desired frequency of oscillation. It is conventional practice to make the three stages of the network identical, so that each provides a 60° phase shift. Although it is common practice, it is also not strictly necessary, provided that the total phase shift is 180°. One reason for using identical stages, however, is that it is possible for the non-identical designs to have more than one frequency for which the total phase shift is 180°. This phenomenon can lead to undesirable multi-modal oscillation.

Figure 10-23C shows the input and output waveforms of an *RC* network in which three stages of 10 kohms/0.01 μF were used with an input frequency of 650 Hz. Note the 180° phase shift in the lower trace. Also, be aware that the vertical input scale factors for these two traces are different. The peak-to-peak amplitude of the upper trace is 7.8 volts, while that of the lower trace is 0.268 volts, or 1/29 of the input amplitude. This attenuation factor is important because it establishes the minimum gain requirement in the amplifier.

Figure 10-24 shows the circuit for an operational amplifier *RC* phase shift oscillator. The cascade phase shift network *R1R2R3/C1C2C3* provides 180° of phase shift at a specific frequency, while the amplifier provides another 180° (because it is an inverting follower). The total phase shift is therefore 360° at the frequency for which the *RC* network provides a 180° phase shift. The frequency of oscillation (*f*) for this circuit is given by:

$$f_{Hz} = \frac{1}{2\pi RC\sqrt{6}} \tag{10-73}$$

where:

f_{Hz} is in hertz (Hz)
R is in ohms (Ω)
C is in farads (F)

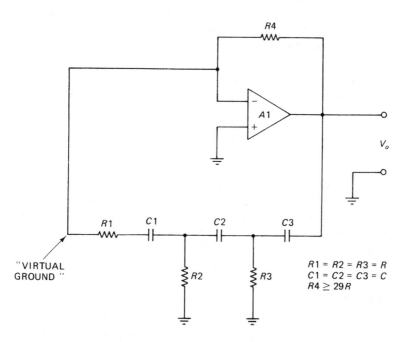

FIGURE 10-24   *RC* phase shift oscillator sinewave generator.

EXAMPLE 10-8

Find the frequency of oscillation for an RC phase shift oscillator (Fig. 10-24) if the three-stage cascade feedback network uses $R = 10$ kohms and $C = 0.005$ μF.

Solution

$$f_{Hz} = \frac{1}{2\pi RC\sqrt{6}}$$

$$f_{Hz} = \frac{1}{(2)(3.14)(\sqrt{6})(10\,000\ \Omega)\left[0.005\ \mu F \times \dfrac{1\ farad}{10^6\ \mu F}\right]}\ Hz$$

$$f_{Hz} = \frac{1}{7.7 \times 10^{-4}}\ Hz = 1300\ Hz \qquad \blacksquare$$

It is common practice to combine the constants in Eq. (10-72) to arrive at a simplified expression:

$$f = \frac{1}{15.39\,RC} \tag{10-74}$$

Because the required frequency of oscillation is usually determined from the application, it is necessary to select an RC time constant to force the oscillator to operate as needed. Also because capacitors come in fewer standard values than resistors, it is common practice to select an arbitrary trial value of capacitance, and then select the resistance that will cause the oscillator to produce the correct frequency. Also, to make the calculations simpler, it is prudent to express the equation in such a way that permits specifying the capacitance (C) in microfarads (μF). As a result Eq. (10-73) is sometimes rewritten as:

$$R = \frac{1\,000\,000}{15.39\,C_{\mu F}\,f} \tag{10-75}$$

The attenuation through the feedback network must be compensated by the amplifier if loop gain is to be unity or greater. At the frequency of oscillation the attenuation is 1/29. The loop gain must be unity, so the gain of amplifier $A1$ must be at least 29 in order to satisfy $A\beta = 1$. For the inverting follower (as shown), $R1 = R$, and $A_v = R4/R1$. Therefore, it can be concluded that $R4 \geq 29R$ in order to meet Barkhausen's criterion for loop gain.

EXAMPLE 10-9

Calculate (a) the resistance required to make an RC phase-shift oscillator operate at a frequency of 1000 Hz when a capacitance of 0.01 μF is used for C in the three-stage cascade phase shift network; and (b) the minimum resistance of $R4$.

Solution

(a) Resistance R required for 1000 Hz:

$$R = \frac{1\,000\,000}{15.39\ C_{\mu F}f_{Hz}}$$

$$R = \frac{1\,000\,000}{(15.39)(0.01\ \mu F)(1000\ Hz)}\ \text{ohms}$$

$$R = \frac{1\,000\,000}{153.9}\ \text{ohms} = 6498\ \text{ohms}$$

(b) Minimum allowable resistance for $R4$:

$$R4 \geq 29\ R$$

$$R4 \geq (29)(6498\ \text{ohms}) = 188\,442\ \text{ohms} \qquad \blacksquare$$

10-7.2 Wien bridge oscillator circuits

The Wien bridge circuit is shown in Fig. 10-25A. Like several other well known bridge circuits, the Wien bridge consists of four impedance arms. Two of the arms ($R1$, $R2$) form a resistive voltage divider that produces an 'output' voltage $V1$ of:

$$V1 = \frac{V_{ac}R2}{R1 + R2} \tag{10-76}$$

The remaining two arms ($Z1$, $Z2$) are complex RC networks that each consist of one capacitor and one resistor. Impedance $Z1$ is a series RC network, while $Z2$ is a parallel RC network. The voltage and phase shift produced by the $Z1/Z2$ voltage divider are functions of the RC values and the applied frequency. Figure 10-25B shows V_{ac} superimposed on $V2$. Note that $V2 = V_{ac}/3$, and that $V2$ and V_{ac} are in-phase with each other.

Figure 10-26 shows the circuit for a Wien bridge oscillator. The resistive voltage divider supplies $V1$ to the inverting input ($-\text{IN}$), while $V2$ is applied to the noninverting input ($+\text{IN}$). In Fig. 10-26 the bridge signal source is the output of the amplifier ($A1$). The AC signal is applied to $+\text{IN}$, so the gain it sees is found from:

$$A_v = \frac{R3}{R4} + 1 \tag{10-77}$$

The AC feedback applied to $+\text{IN}$ is:

$$\beta = \frac{Z2}{Z1 + Z2} \tag{10-78}$$

At resonance $\beta = 1/3$, so (as shown in Fig. 10-25B):

$$V2 = \frac{V_o}{3} \tag{10-79}$$

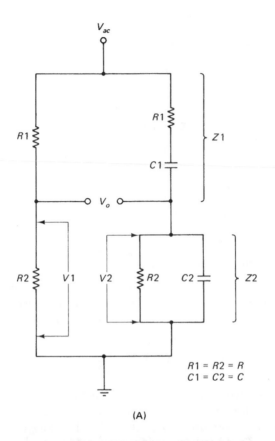

$$R1 = R2 = R$$
$$C1 = C2 = C$$

(A)

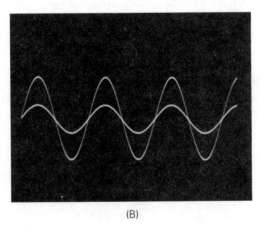

(B)

FIGURE 10-25   (A) Wien bridge network; (B) input and output waveforms.

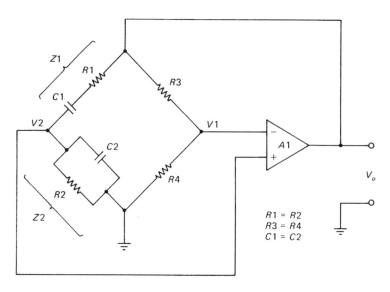

FIGURE 10-26 Wien bridge oscillator circuit.

Because $A_v = V_o/V_2$ by definition, satisfying Barkhausen's loop gain criterion ($A_v\beta = 1$) requires that $A_v = V_o/V_2 = 3$. Using this result:

$$A_v = \frac{R3}{R4} + 1 \qquad (10\text{-}80)$$

$$3 = \frac{R3}{R4} + 1 \qquad (10\text{-}81)$$

$$2 = \frac{R3}{R4} \qquad (10\text{-}82)$$

or

$$R3 = 2\,R4 \qquad (10\text{-}83)$$

If $R1 = R2 = R$ and $C1 = C2 = C$, the resonant frequency of the Wien bridge is:

$$f_{Hz} = \frac{1}{2\pi RC} \qquad (10\text{-}84)$$

For the standard Wien bridge oscillator, in which $R1 = R2 = R$ and $C1 = C2 = C$, and $R3 = 2\,R4$, a sinewave output will be produced with frequency f_{Hz}.

EXAMPLE 10-10

Calculate the resistor values for a Wien bridge oscillator which produces a frequency of 1000 Hz.

Solution

(a) Let $C = 0.01$ μF (trial value) and $R4 = 10$ kohms.

(b) Set gain resistor for $A_v = 3$: $R3 = 2 R4 = (2)(10$ kohms$) = 20$ kohms.

(c) Set $R1 = R2 = R$ value.

(d) The design values are, therefore:

$$R = \frac{1}{2\pi RC}$$

$$R = \frac{1}{(2)(3.14)(1000 \text{ Hz}) \left[0.01 \text{ μF} \times \dfrac{1 \text{ F}}{10^6 \text{ μF}}\right]}$$

$$R = \frac{1}{6.28 \times 10^{-5}} = 15\,924 \text{ ohms}$$

$R1, R2$: 15 924 ohms

$R3$: 20 kohms

$R4$: 10 kohms

$C1, C2$: 0.01 μF

f: 1000 Hz ■

Amplitude stability. The oscillations in the Wien bridge oscillator circuit want to build up without limit when the gain of the amplifier is high. Figure 10-27 shows the result of the gain being only slightly above that required for stable oscillation. Note that some clipping is beginning to appear on the sinewave peaks. At even higher gains the clipping becomes more severe, and will eventually look like a squarewave. Figure 10-28 shows several methods for stabilizing the waveform amplitude. Figure 10-28A

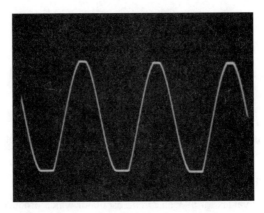

FIGURE 10-27   Output clipping of normal Wien bridge oscillator.

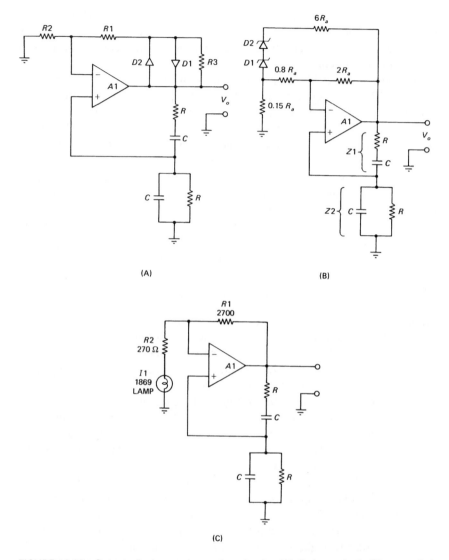

FIGURE 10-28 Output clipping compensation circuits: (A) diode method; (B) zener diode method; (C) lamp method.

shows the use of small-signal diodes such as the 1N914 and 1N4148 devices. At low signal amplitudes the diodes are not sufficiently biased, so the gain of the circuit is:

$$A_v = \frac{R1 + R3}{R2} + 1 \qquad (10\text{-}85)$$

As the output signal voltage increases, however, the diodes become forward biased. Diode $D1$ is forward biased on negative peaks of the signal, while diode $D2$ is forward biased on positive peaks. Because $D1$ and $D2$ are shunted across $R3$, the total resistance $R3'$ is less than $R3$. By inspection of Eq. (10-82) one can determine that reducing $R3$ to $R3'$ reduces the gain of the circuit. The circuit is thus self-limiting.

Another variant of the gain-stabilized Wien bridge oscillator is shown in Fig. 10-28B. In this circuit a pair of back-to-back zener diodes provide the gain limitation function. With the resistor ratios shown the overall gain is limited to slightly more than unity, so the circuit will oscillate. The output peak voltage of this circuit is set by the zener voltages of $D1$ and $D2$ (which should be equal for low-distortion operation).

One final version of the gain stabilized oscillator is shown in Fig. 10-28C. In this circuit a small incandescent lamp is connected in series with resistor $R2$. When the amplitude of the output signal tries to increase above a certain level, the lamp will draw more current causing the gain to reduce. The lamp-stabilized circuit is probably the most popular form where stable outputs are required. A thermistor is sometimes substituted for the lamp.

10-7.3 Quadrature and biphasic oscillators

Signals that are *in quadrature* are of the same frequency but are phase shifted $90°$ with respect to each other. An example of quadrature signals is sine and cosine waves (Fig. 10-29A). Applications for the quadrature oscillator include demodulation of phase sensitive detector signals in data acquisition systems. The sinewave has an instantaneous voltage $v = V \sin(\omega_0 t)$, while the cosine wave is defined by $v = V \cos(\omega_0 t)$. Note that the distinction between sine and cosine waves is meaningless unless either both are present, or some other timing method is used to establish when 'zero degrees' is supposed to occur. Thus, when sine and cosine waves are called for it is in the context of both being present, and a phase shift of $90°$ is between them.

The circuit for the quadrature oscillator is shown in Fig. 10-29B. It consists of two operational amplifiers, $A1$ and $A2$. Both amplifiers are connected as Miller integrators, although $A1$ is a noninverting type while $A2$ is an inverting integrator. The output of $A1$ (i.e., V_{o1}) is assumed to be the sinewave output. In order to make this circuit operate, a total of $360°$ of phase shift is required between the output of $A1$, around the loop and back to the input of $A1$. Of the required $360°$ phase shift $180°$ are provided by the inversion inherent in the design of $A2$ (it is in the inverting configuration). Another $90°$ obtains from the fact that $A2$ is an integrator, which inherently causes a $90°$ phase shift. An additional $90°$ phase shift is provided by RC network $R3C3$. If $R1 = R2 = R3 = R$, and $C1 = C2 = C3 = C$, then the frequency of oscillation is given by:

$$f_{\text{Hz}} = \frac{1}{2\pi RC} \tag{10-86}$$

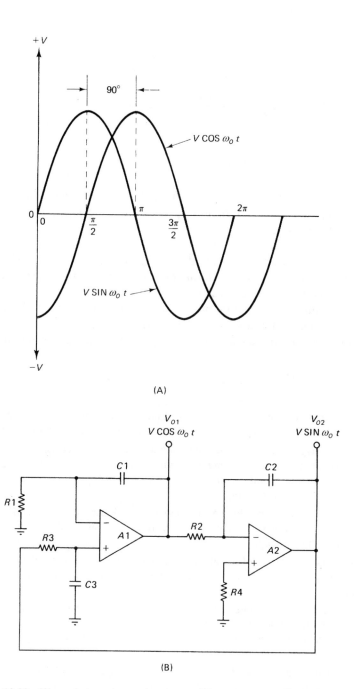

FIGURE 10-29 (A) quadrature sine cosine waves; (B) quadrature oscillator circuit.

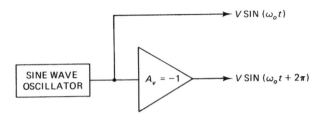

FIGURE 10-30 Biphasic oscillator.

The cosine output (V_{o2}) is taken from the output of amplifier $A2$. The relative amplitudes are approximately equal, but the phase is shifted $90°$ between the two stages.

A *biphasic oscillator* is a sinewave oscillator that outputs two identical sinewave signals that are $180°$ out of phase with each other. The basic circuit is simple, and is shown in block diagram form in Fig. 10-30. The biphasic oscillator consists of a sinewave oscillator followed by an inverting amplifier that has a gain of one. The output of the sinewave oscillator is $V \sin(\omega_0 t)$, while the output of the inverter is $V \sin(\omega_0 t + 2\pi)$. Biphasic oscillators are sometimes used in transducer excitation applications in carrier amplifiers.

10-8 SUMMARY

1. Monostable and astable multivibrators are based on the properties of the operational amplifier connected as a voltage comparator. In both cases timing of the output waveform is based on the properties of the simple series RC network.

2. A nonretriggerable monostable multivibrator cannot be retriggered until both the output pulse duration and any refractory period expire. A retriggerable monostable multivibrator will respond to additional trigger pulses regardless of time-out or refractory period.

3. The sawtooth and triangle generator circuits are designed using the Miller integrator as a ramp generator. When driven from a constant voltage the integrator outputs a linear ramp.

4. The principle causes of frequency and amplitude stability problems in fixed frequency sinewave oscillators are *temperature* and *power supply voltage variation*.

5. RC phase shift oscillators use the amplifier to provide $180°$ of phase shift, and a three-section cascade RC network to provide the remaining $180°$ required to meet Barkhausen's criterion. The amplifier must provide a voltage gain of 29 or greater.

6. The Wien bridge oscillator uses a bridge consisting of a parallel and a series RC network, along with a resistor voltage divider in an operational amplifier circuit.

7. Amplitude stability is provided in the Wien bridge oscillator by one of several means, a diode AGC, a zener diode AGC, and an incandescent lamp in series with one of the arms.

8. The quadrature oscillator is built using an inverting integrator, a noninverting integrator, and an *RC* phase shift network to create the required 360° phase shift.

10-9 RECAPITULATION

Now return to the objectives and Pre-quiz questions at the beginning of the chapter and see how well you can answer them. If you cannot answer certain questions, mark them and review the appropriate parts of the text. Next, try to answer the questions and work the problems below using the same procedure.

10-10 STUDENT EXERCISES

1. Design and build a one-shot multivibrator which has an output duration of 5 ms. Examine the output voltage and capacitor voltage with an oscilloscope. Drive the one-shot with a pulse generator or squarewave generator and then compare the output pulse and trigger signal.

2. Design and build a symmetrical squarewave generator operating at a frequency of 1000 Hz using equal positive feedback resistors. Examine the output and capacitor voltages on an oscilloscope. Vary the values of *R*1 and *C*1, and observe what happens to the circuit. Replace the positive feedback resistors with a single potentiometer. Initially set the potentiometer to midscale. Observe what happens to the operating frequency and waveform when the ratio of resistances is changed by varying the potentiometer.

3. Build an asymmetrical squarewave oscillator using (a) potentiometer method, (b) diode method. In both cases examine operation of the circuit on an oscilloscope.

4. Design and build a triangle waveform oscillator operating on a frequency of 500 Hz.

5. Design and build a sawtooth oscillator operating on a frequency of 100 Hz.

6. Design and build an *RC* phase shift oscillator operating on a frequency of 1000 Hz. Use a 741 or other available operational amplifier. Use an oscilloscope to check the output voltage and the feedback voltage. Confirm the feedback attenuation is 1/29. Vary the *RC* network values and see what happens to the circuit.

7. Design and build a Wien bridge oscillator operating on a frequency of 600 Hz. Use an oscilloscope to examine the output signal. Confirm the feedback attenuation factor is 1/3. Vary the component values and observe what happens in the circuit. Redesign the circuit using one or more of the amplitude stability techniques.

10-11 QUESTIONS AND PROBLEMS

1. Calculate the *RC* time constant that will allow a capacitor to charge from −12 Vdc to +10 Vdc in 350 ms.

2. An op-amp monostable multivibrator (MMV) must produce a 70 ms pulse. Select an *RC* time constant.

3. A monostable multivibrator (Fig. 10-4A) is constructed with $R2 = 10$ kohms and $R3 = 4.7$ kohms. The device is a 741 op-amp operated at $|V_{sat}| = 12$ Vdc. Calculate the RC time constant required to produce an 8 ms output pulse.

4. A monostable multivibrator has a timing resistor of 9.1 kohms and a timing capacitor of 0.01 μF. What is the duration of the output pulse if the positive feedback resistors are equal?

5. Calculate the duration of a retriggerable monostable multivibrator circuit in which $R2 = 10$ kohms, $R3 = 8.2$ kohms, $R1 = 68$ kohms and $C1 = 0.068$ μF.

6. Calculate the oscillating frequency for an astable multivibrator in which $R1 = 10$ kohms, $R2 = 10$ kohms, $R3 = 5.6$ kohms, and $C1 = 0.01$ μF.

7. An asymmetrical astable multivibrator is designed such that $R1A = 82$ kohms, $R1B = 22$ kohms, $C1 = 0.08$ μF and $R2 = R3$.

8. Calculate the frequency of an astable multivibrator in which the positive feedback resistors are equal and the RC time constant is 0.010 seconds.

9. State Barkhausen's criteria for oscillation.

10. Calculate the oscillating frequency when an RC phase shift oscillator has the following component values: $R1 = R2 = R3 = 15$ kohms, $C1 = C2 = C3 = 0.1$ μF. Calculate the minimum value of gain setting feedback resistor $R4$.

11. Calculate the resonant frequency of a Wien bridge oscillator in which $R = 100$ kohms and $C = 0.001$ μF.

12. Find the frequency of oscillation for an RC phase shift oscillator (Fig. 10-24) if the three-stage cascade feedback network uses $R = 22$ kohms and $C = 0.002$ μF.

13. Describe the operation of a relaxation oscillator. Give two examples of electronic devices that can be used in making relaxation oscillators.

14. A series RC network consists of a 1.5 μF capacitor and a 2.2 megohm resistor. The network is connected to a 20 volt DC power supply for 20 minutes. The capacitor is charged to a potential of _____ volts DC. Assume that the DC power supply is replaced with a short circuit, and calculate: (a) the length of time for the capacitor to discharge to 7.34 volts; (b) the potential across the capacitor after 3.3 seconds; and the time required for the potential across the capacitor to reach zero.

15. A series RC network consists of a 0.47 μF capacitor and an 82 kohm resistor. How much time is required for the capacitor to charge from $+10$ Vdc to -2 Vdc?

16. What is the difference between a monostable and an astable multivibrator'?

17. What is the principal difference between a retriggerable one-shot and a nonretriggerable one-shot? Use waveform drawings if necessary to make your point.

18. What are the four states of a monostable multivibrator'?

19. Draw the block diagram for a feedback oscillator.

20. What is the rule-of-thumb maximum value of the time constant $R4C2$ in Fig. 10-4A if $R1C1 = 10$ ms?

21. In Fig. 10-4A, what is the value of $V1$ when $V_{o,} = -12$ Vdc and $R2 = R3$?

22. The circuit of Fig. 10-4A is stable. Calculate differential input voltage V_{id}, if $R2 = R3$ and the maximum allowable output voltage is $+9$ Vdc.

23. Describe the *refractory period* of a monostable multivibrator.

24. What is the function of $D3/R5$ in Fig. 10-4A? What is the effect of these parts on the V_c waveform in Fig. 10-4B?

25. Draw the circuit for a retriggerable multivibrator.

26. Draw the circuit for an operational amplifier astable multivibrator. Calculate values for the timing components for a 1000 Hz version of this circuit when the feedback factor (β) is 0.5.

27. Draw the timing diagram for the circuit of question 26. Label the time durations of each section.

28. Draw the schematic diagram for an astable multivibrator that has a continuously adjustable duty cycle.

29. Why does the circuit of Fig. 10-11A use two timing resistors ($R1A$ and $R1B$)? What is the function of diodes $D1$ and $D2$?

30. In Fig. 10-13A the zener potential of $D1$ is 5.6 Vdc, and of $D2$ is 6.8 Vdc. Calculate the maximum output voltages of both positive and negative excursions of the output waveform.

31. Discuss a simple means for making a ramp generator circuit from a Miller integrator.

32. Sketch triangle and sawtooth waveforms and describe the difference between them.

33. Draw the constant current and normal RC charging curves for a Miller integrator.

34. Draw the circuit for a simple RC sawtooth generator. Also for an op-amp sawtooth generator based on the Miller integrator.

35. What is a staircase generator and how does it differ from a sawtooth generator?

36. Draw the frequency determining network for an RC phase shift oscillator. What is the total phase shift across this network at the resonant frequency? What is the approximate voltage gain/loss of the network?

37. What is the minimum value of $R4$ in Fig. 10-24 if $R1 = R2 = R3 = 2.7$ kohms?

38. What is the purpose of $D1$ and $D2$ in Fig. 10-28A?

39. Why is the incandescent lamp used in Fig. 10-28C?

40. The input signal to an op-amp circuit is $v = V \sin(\omega_0 t)$. Write the output signal expressions if the circuit is (a) an inverting follower with a gain of one, (b) an inverting follower with a gain of ten, (c) a noninverting follower with a gain of two, (d) an integrator, and (e) a differentiator.

CHAPTER 11

DC power supplies for linear IC circuits

OBJECTIVES

1. Learn the basic components required to make a low-voltage DC power supply.
2. Learn the required ratings, specifications and safety margins for DC power supply circuits.
3. Be able to design common low-voltage DC power supplies.

11-1 PRE-QUIZ

These questions test your prior knowledge of the material in this chapter. Try answering them before you read the chapter. Look for the answers (especially those you answered incorrectly) as you read the text. After you have finished studying the chapter try answering these questions again, and those at the end of the chapter (see Section 11-16).

1. State the maximum operating current and output potential of an LM-340T-12 three-terminal IC voltage regulator.
2. A zener diode regulator must regulate a DC supply to +6.8 Vdc. The input DC source varies from 9 to 13 volts. Assume a constant load current of 90 mA. Find the value of series resistor, its power dissipation, and the power dissipation of the zener diode.
3. A 'brute force' ripple filter in a 12 Vdc power supply is required to reduce the fullwave ripple to 0.8 when the load current is 1.2 amperes. Calculate the value of filter capacitance needed.

4. A 12 Vdc power supply must deliver 900 mA approximately 50% of the time. What package should be specified for a three-terminal IC regulator that will do this job?

11-2 INTRODUCTION TO DC POWER SUPPLIES ─────────

The DC power supply is important to the success of any electronic circuit or equipment. The power supply converts the alternating current available from the power mains to the direct current needed to operate electronic circuits. The typical DC power supply consists of several different components: *transformer, rectifier, ripple filter*, and (in some designs) *voltage regulator*. The transformer scales the AC voltage from the power lines up or down as needed for the particular application. The job of the rectifier is to convert the bidirectional AC into unidirectional *pulsating DC*, while the ripple filter smooths the pulsating DC into nearly pure-DC. The voltage regulator is used to stabilize the voltage in the face of changing load currents and AC input voltage.

Also part of some DC power supplies are functions such as *overvoltage protection*, and *current-limiting*. These circuits, as well as the main components of the DC power supply are discussed in detail in this chapter. You will learn the fundamentals of DC power supply design, especially the regulated, low-voltage DC power supplies that are typically used with circuits containing linear integrated circuit elements.

11-3 RECTIFIERS ──────────────────────────────

Rect-i-fy: 'To make right; remove impurities'. The purpose of a rectifier in a DC power supply circuit is to remove the impurities of the AC line current and make it right for DC-craving electronic circuits that require pure, or nearly pure, direct current.

Before discussing the details of solid-state rectifiers, let's review the two basic forms of electrical current in the context of rectification: DC and AC (see Fig. 11-1). Direct current (DC) is graphed in Fig. 11-1A. The key feature of this form of electrical current is that it is *unidirectional*, i.e. current flows through the circuit in only one direction. It will be zero until turned on (time $T1$), and will then rise to a certain level and remain there. Electrons flow from the negative terminal to the positive terminal of the power supply, and that polarity never reverses direction.

Alternating current, on the other hand, is *bidirectional* (see Fig. 11-1B). On one halfcycle the current flows in one direction. Then the power supply polarity reverses, so the current flows in the opposite direction. The electrons still flow from negative to positive, but since the positive and negative poles have switched places, the physical direction of current flow has reversed. In the normal AC power mains the voltage and current waveforms vary as a

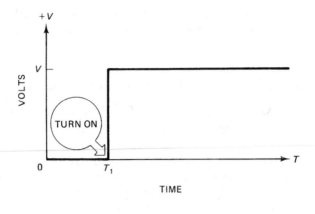

(A)

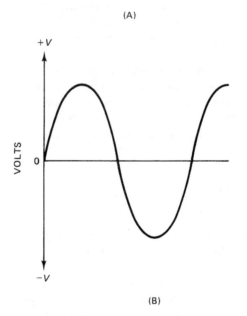

(B)

FIGURE 11-1   (A) DC current; (B) AC current.

sinewave. By convention, flow in the 'positive' direction is graphed above the zero volts (or zero amperes) line, and flow in the 'negative' direction is graphed below the zero line.

AC is incompatible with nearly all electronic circuits, so must be changed to DC by a rectifier and a ripple filter. The main requirement for a rectifier is that it convert bidirectional AC into a unidirectional form of current. Although industry once used rotary mechanical switches, synchronous vibrators, and vacuum tubes to accomplish the rectification job, all modern circuits rely on solid-state PN junction rectifiers.

11-3.1 PN junction diode rectifiers

The modern solid-state rectifier really isn't so new after all. Various versions of the rectifier date back to the dawn of the radio — indeed the electrical — age. All common rectifier diodes in use today, however, are silicon PN junction diodes (shown schematically in Fig. 11-2).

The PN junction diode rectifier (Fig. 11-2) consists of a silicon semiconductor material that is doped with impurities to form N-type material at one end and P-type material at the other end. The charge carriers (which form the electrical current) in the N-type material are negatively charged electrons, while the charge carriers in the P-type material are positively charged 'holes'.

The reverse bias situation is shown in Fig. 11-2A. In this case the negative terminal of the voltage source (V) is connected to the P-type material, while the positive terminal is connected to the N-type material. Positive charge carriers are thus attracted away from the PN junction towards the negative voltage terminal, while negative charge carriers are drawn away towards the positive terminal. That leaves a charge-free *depletion zone* in the region of the junction that contains no carriers. Under this condition, there is little or no current flow across the junction. Theoretically, the junction current is zero, although in real diodes there is always a tiny *leakage current* across the junction.

The forward biased case is shown in Fig. 11-2B. Here the polarity of voltage source V is reversed from Fig. 11-2A. The positive terminal is applied to the P-type material, and the negative terminal is applied to the N-type material. Because like charges repel, the charge carriers in both P-type and N-type material are driven away from the power supply terminals towards the junction. The depletion zone disappears, allowing positive and negative charges to get close to the boundary between regions. As these opposite charges attract each other across the junction, a current flows in the circuit.

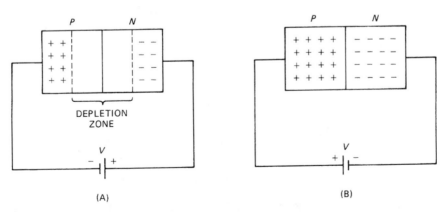

FIGURE 11-2   PN junction diode: (A) reverse biased; (B) forward biased.

From the above description it is apparent that a PN junction diode is able to convert bidirectional AC into unidirectional current because it allows current to flow in only one direction. Thus, it is a rectifier. However, the rectifier output current is not pure DC (as from batteries), but rather it is pulsating DC.

Figure 11-3 shows the standard circuit symbol for the solid-state rectifier diode (Fig. 11-3A), along with some common shapes of actual diodes. The 'input' side where AC is applied is the anode, while the 'DC' output is the cathode. The diodes shown in Figs 11-3B through 11-3G are positioned so that the respective anodes and cathodes are aligned with those of the circuit symbol in Fig. 11-3A. Rectifiers 11-3B through 11-3E are epoxy package devices, and are the type seen most often. The cathode end will be marked either with a rounded end (11-3B), a line (11-3C), a diode arrow (11-3D) or a plus sign (11-3E).

The diode shown in Fig. 11-3F is the old-fashioned (now obsolete) 'tophat' type. Unless otherwise specified, the tophat type can safely pass a current of 500 milliamperes, while those in Figs 11-3B through 11-3E generally pass 1 ampere (or more, for larger sized but similar packages).

The stud-mounted type shown in Fig. 11-3G is a high current model. These diodes are rated at currents from 6 amperes and up. (50 and 100 ampere models are easily obtained). These diodes are mounted using a threaded screw at one end, which also forms one electrical connection. The other electrical connection is the solder terminal at the other end. Unless otherwise specified, the solder terminal is the anode, while the stud-mount is the cathode terminal. Exceptions to the polarity rule are sometimes seen. The reverse polarity diodes will have either an arrow symbol pointing in the opposite direction (the arrow always points to the cathode), or an 'R' suffix on the type number (e.g. 1NxxxxR instead of 1Nxxxx).

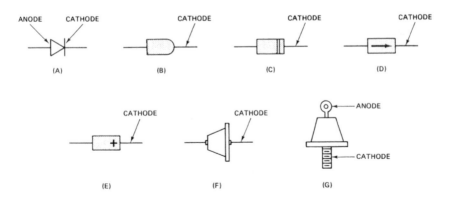

FIGURE 11-3   (A) rectifier circuit symbol; (B) thru (E) epoxy-plastic case rectifier diodes; (F) 'tophat' rectifier diode; (G) stud-mounted high current diode.

11-3.2   Rectifier specifications

The proper use of solid-state rectifiers requires consideration of several key specifications: *forward current, leakage current, surge current, junction temperature* and *peak inverse voltage* (PIV) — also called *peak reverse voltage* (PRV).

The forward current is the maximum constant current that the diode can pass without damage. For the 1N400x series of rectifiers the forward specified current is one ampere (1 A). The leakage current is the maximum current that will flow through a reverse biased junction. In an ideal diode, the leakage current is zero while in high quality practical diodes it is very low. The surge current is typically very much larger than the forward current, and it is sometimes erroneously taken to be the operating current of the diode. Surge current is defined as the maximum short duration current that will not damage the diode. 'Short duration' typically means one AC cycle (1/60 second in a 60 Hz system, 1/50 second in a 50 Hz system). Don't use the surge current as if it were the forward current.

The specified junction temperature is the maximum allowable operating temperature of the PN junction. The actual junction temperature depends upon the forward current and how well the package (and environment) rids the diode of internal heat. Although typical maximum junction temperatures range up to $+125°C$, good design requires as low a temperature as possible. One reliability guide requires that the junction temperature be held to a maximum of $+110°C$.

The peak inverse voltage (PIV) is the maximum allowable reverse bias voltage that will not damage the diode. This rating is usually the limiting rating in certain power supply designs.

11-3.3   Rectifier circuits

Figure 11-4 shows a solid-state rectifier diode (*D*1) in a simple *halfwave rectifier* circuit. In Fig. 11-4A the diode is forward biased: the positive terminal of the voltage source is connected to the anode of the rectifier. Current (*I*) flows through the load resistance (*R*). In Fig. 11-4B the opposite situation is found: the negative terminal of the voltage source is applied to the anode, so the diode is reverse biased and no current flows.

The circuit of Fig. 11-4 is called a halfwave rectifier for reasons that become apparent when examining Fig. 11-4C. In this figure the output current through the load (*R*) is graphed as a function of time when an AC sinewave is applied. From time *T*1 to *T*2 the diode is forward biased, so current flows in the load (also from *T*3 to *T*4). But during the period *T*2 to *T*3 the diode is reverse biased, so no current flows. Because the entire sinewave takes up the period *T*1 to *T*3, only half of the input sinewave is used. The output waveform shown in Fig. 11-4C is called a *halfwave rectified pulsating DC wave*.

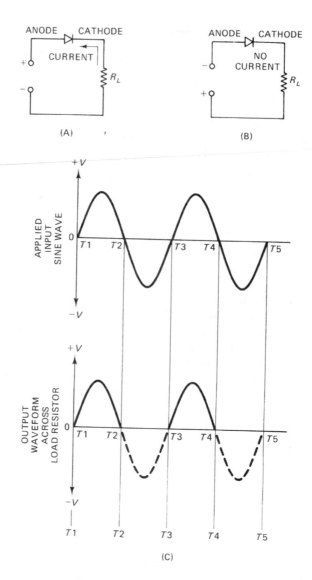

FIGURE 11-4 Halfwave rectifier circuit: (A) forward biased; (B) reverse biased; (C) input and output waveforms.

The halfwave rectifier is low-cost, but wastes energy due to its use of only one-half of the input AC waveform. Efficiency is increased by making use of the entire waveform in a *fullwave rectifier circuit*. Figure 11-5A shows the standard fullwave rectifier. This circuit uses a transformer that has a center-tapped secondary winding. Because the center tap (CT) is used as the zero volts reference (and in most circuits is grounded), the polarities at

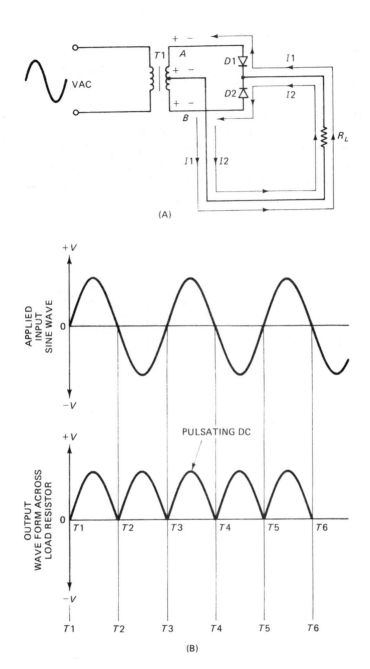

FIGURE 11-5 (A) Standard fullwave rectifier circuit; (B) input and output waveforms.

the ends of the secondary are always opposite each other (i.e. 180°). On one halfcycle, point A is positive with respect to the CT, while point B is negative. On the next halfcycle, point A is negative and point B is positive with respect to the CT. This situation makes *D1* forward biased on one halfcycle, while *D2* is reverse biased. Alternatively, on the next halfcycle, *D1* is reverse biased and *D2* is forward biased.

Follow the circuit of Fig. 11-5A through one complete AC cycle (times *T1* through *T3* in Fig. 11-5B). On the first halfcycle (*T1–T2*), point A is positive, so *D1* is forward biased and conducts current; *D2* is reverse biased. Current *I1* flows from the CT, through load *R*, diode *D1* and back to the

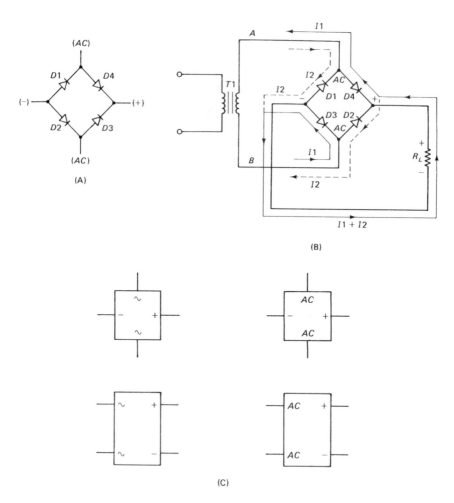

(A)

(B)

(C)

FIGURE 11-6 (A) Bridge rectifier circuits; (B) current flows in bridge rectifier circuit; (C) schematic circuit symbols.

transformer at point A. On the alternate halfcycle, current $I2$ flows from the CT, through load R, diode $D2$ and back to the transformer at point B. Now notice what happened: $I1$ and $I2$ are equal currents, generated on alternate halfcycles, and they *flow in load R in the same direction*. Thus, a unidirectional current is flowing through load R on both halves of the AC sinewave. The waveform resulting from this action is shown in Fig. 11-5B, and is called *fullwave rectified pulsating DC*.

The center tap on the secondary can be eliminated by using the fullwave bridge rectifier circuit of Fig. 11-6A. This circuit requires twice as many rectifier diodes as the other form of fullwave rectifier, but allows the use of a simpler transformer (no center tap). The operation, however, is similar (see Fig. 11-6B). On one halfcycle, point A is positive and point B is negative. Current $I1$ flows from the transformer at point B, through $D4$, load R, diode $D1$ and back to the transformer at point A. On the alternate halfcycle, point A is negative and point B is positive. In this case, current $I2$ flows from point A, through diode $D3$, load R (in the same direction as $I1$), diode $D2$ and back to the transformer at point B.

A bridge rectifier can be built using four discrete diodes ($D1-D4$). In most modern equipment, however, a bridge stack is used. These parts are a bridge rectifier built into a single package with four leads coming out. Figure 11-6C shows various alternative circuit symbols for bridge rectifier stacks.

Figure 11-7 shows a 'halfbridge' fullwave rectifier. This circuit is used more today because of the dual polarity power supplies used in a lot of equipment. Operational amplifiers and some CMOS devices typically require ± 12 volt DC power supplies. A fullwave bridge rectifier stack coupled to a center-tapped transformer will make a pair of fullwave rectified DC power

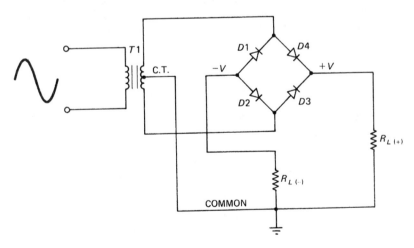

FIGURE 11-7 Dual polarity fullwave rectifier uses center-tapped transformer to establish two different output polarities.

supplies. The CT is the common (or ground), while the bridge positive terminal supplies the positive voltage and the negative terminal supplies the negative voltage. Both outputs from the halfbridge rectifier are fullwave rectified pulsating DC.

11-3.4 Selecting rectifier diodes

The two parameters most often used to specify practical power supply diodes are the forward current and peak inverse voltage.

The forward current rating of the diode must be at least equal to the maximum current load that the power supply must deliver. In practical circuits there is also a need for a safety margin to account for tolerances in the diodes and variations of the real load (as opposed to the calculated load). It is also true that making the rating of the diode larger than the load current will greatly improve reliability. A good rule of thumb is to select a diode with a forward current rating of 1.5 to 2 times the calculated (or design goal) load current — or more if available. Selecting a diode with a very much larger forward current (e.g. 100 amperes for a 1 ampere circuit) is both wasteful and likely to make the diode not work exactly like a rectifier diode. The general rule is to make the rating as high as feasible. The 1.5 to 2 times rule, however, should result in a reasonable margin of safety.

The peak inverse voltage (PIV) rating can be a little more complicated. In unfiltered, purely resistive, circuits the PIV rating need only be greater than the maximum peak applied AC voltage (1.414 times RMS). But if a 20% safety margin is desired, then make it 1.7 times RMS voltage.

Most rectifiers are used in *ripple filtered* circuits (e.g. Fig. 11-8A), and that makes the problem different. Figure 11-8B shows the simple halfwave rectifier capacitor filtered circuit redrawn to better illustrate the circuit action. Keep in mind that capacitor $C1$ is charged to the peak voltage with the polarity shown. This peak voltage is 1.414 times the RMS voltage. The peak

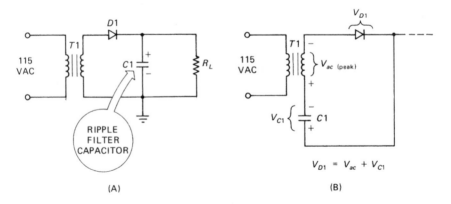

FIGURE 11-8   (A) 'Brute force' ripple filter; (B) circuit redrawn.

voltage across the transformer secondary (V) is in series with the capacitor voltage. When voltage V is positive, the transformer voltage and capacitor voltage cancel out, making the diode reverse voltage nearly zero. But when the transformer voltage (V) is negative, the two negative voltages (V and V_c) add up to twice the peak voltage:

$$V_{D1} = V_{ac(peak)} + V_{C1} \tag{11-1}$$

but,

$$V_{C1} = V_{ac(peak)} \tag{11-2}$$

so,

$$V_{D1} = V_{ac(peak)} + V_{ac(peak)} \tag{11-3}$$

$$V_{D1} = 2V_{ac(peak)} \tag{11-4}$$

Because $V_{ac(peak)} = 1.414 V_{ac(rms)}$:

$$V_{D1} = 2(1.414 V_{ac(rms)}) \tag{11-5}$$

$$V_{D1} = 2.828 V_{ac(rms)} \tag{11-6}$$

The reverse voltage across the diode is approximately 2.83 times the RMS voltage. Therefore, the absolute minimum value of PIV rating for the diode is 2.83 times the applied RMS. If a 20% safety margin is preferred, then the diode PIV rating should be 3.4 times the applied RMS voltage (or more).

11-3.5   Using rectifier diodes

In most cases, especially low-voltage power supplies, diodes can be used as shown in the circuits above. In Fig. 11-9A, however, the proper way to use the solid-state diode rectifier is shown. The resistor in series with diode $D1$ (i.e. $R1$) is used to limit the forward current. Many circuits, especially those with capacitor-input filter circuits, exhibit a large surge current at initial turn-on. This current can sometimes destroy the diode, so $R1$ is used to limit the possible damage. The resistance value of $R1$ is typically 5 to 20 ohms. In most cases, however, $R1$ can be eliminated by using a diode with a current rating significantly larger than the load current (for example, the two times rule). Also, the transformer secondary resistance (R_s) serves the current-limiting function of $R1$ in many circuits.

Capacitor $C1$ in Fig. 11-9A is used to bypass high voltage transient spikes around the diode. These spikes could possibly blow the diode PN junction. In fact, high voltage line spikes are a frequent source of damage to rectifier diodes. Placing the capacitor in parallel with the diode will eliminate that problem. The DC working voltage (WVDC) of the capacitor should be equal to or greater than the PIV rating of the diode.

By use of 1000-volt PIV diodes (even in low-voltage circuits) much of the damage caused by transients is avoided. The capacitors can often be

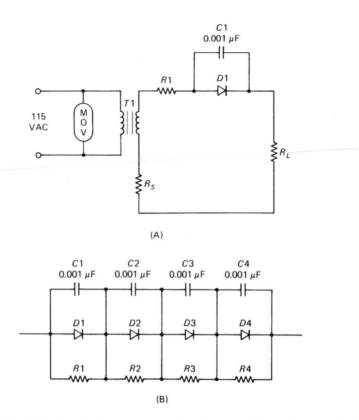

FIGURE 11-9 (A) Diode protection measures; (B) using several diodes in series for high voltage.

eliminated if a *metal oxide varistor* (MOV) or some other high voltage spike suppressor device is shunted across the AC supply voltage in the primary circuit of the power transformer (see Section 11-12).

Figure 11-9B shows the method for connecting several diodes in series to increase the PIV rating. Assuming that the PIV ratings of the diodes are equal, then the overall rating is four times the rating of one diode. If 1000-volt PIV diodes are used in this circuit, the total PIV rating of the assembly is 4000 volts.

The capacitors used in Fig. 11-9B are for exactly the same purpose as in Fig. 11-9A. The resistors, however, are needed for a different purpose: they equalize the forward voltage drop across each diode. A 470 kohm, 2 watt resistor is typically used for 1000-volt PIV diodes. The 2 watt rating is required not because of the power dissipation of the resistors, but rather for their voltage rating (yes, resistors do have voltage ratings).

Figure 11-10 shows the proper method for mounting an axial lead rectifier on a prototyping 'perfboard' or printed circuit board. This method is used

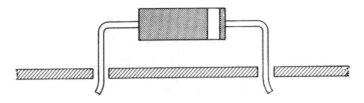

FIGURE 11-10   Mounting to cool rectifier diode where vibration is not a serious factor.

anytime except where excessive vibration is expected, which includes most sedentary circuits or equipment. The space beneath the diode body allows air to circulate (keeping the diode cooler) and prevents diode heat from damaging the board.

11-4   RIPPLE FILTER CIRCUITS

The pulsating DC output from either fullwave or halfwave rectifiers is almost as useless for some electronic circuits as the AC input waveform. A ripple filter circuit is used to smooth out the pulsating DC into a

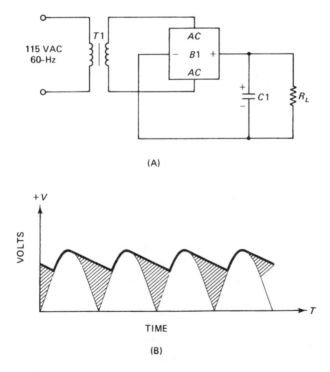

(A)

(B)

FIGURE 11-11   (A) Fullwave bridge rectifier circuit with ripple filter; (B) output waveform showing action of filter capacitor.

purer DC. Figure 11-11A shows the simplest form of filter circuit: a single capacitor ($C1$) connected in parallel with the load. Circuit action is shown in Fig. 11-11B. The job of the capacitor is to store electrical charge on voltage peaks, and then dump that charge into the load when the voltage drops between peaks. The shaded area of Fig. 11-11B shows the filling in caused by the capacitor charge. The output voltage is the sum of both rectifier and capacitor contributions, and is represented by the heavy line in Fig. 11-11B.

The value of the filter capacitor is determined by the amount of ripple factor that can be accepted. The ripple factor (r) is defined as the ratio of the ripple voltage amplitude to the average voltage of the rectified waveform. Values tend to be in the range 0.01 to 1.0 for common electronic circuits. The 'rule of thumb' for ripple factor for 60 Hz circuits is:

Halfwave rectified circuits:

$$r = \frac{1\,000\,000}{208\ C1R_L} \tag{11-7}$$

Fullwave rectified circuits:

$$r = \frac{1\,000\,000}{416\ C1R_L} \tag{11-8}$$

where $C1$ is the capacitor value in microfarads and R_L is the load resistance in ohms. The load resistance is the ratio of the output voltage to the output current: V_o/I_o.

So what do these figures mean in practical terms? A few oscilloscope waveform photos (Fig. 11-12) illustrate the effect of adding capacitance to the filter circuit. In all cases the AC applied to the rectifier was nominally 12 VAC, and the load resistance was 25 ohms. These photos were taken with the 'scope AC-coupled to permit expansion of the ripple in the presence of the large DC offset. The waveforms shown in Fig. 11-12 ride on top of the DC output applied to the load, but the DC offset is suppressed by the 'scope input circuit. The 'scope sensitivity was 5 V/cm.

The unfiltered, fullwave rectified pulsating DC (Fig. 11-12A) was measured with a DC voltmeter and found to be 9.25 Vdc. This value is not the peak voltage (which was close to 12 Vdc), but rather an average value caused by the fact that the DC meter tends to average the reading. When a 150 µF capacitor is connected across the 25 ohm load the ripple reduces to 0.64 (Fig. 11-12B), and the output voltage rises to 10.86 Vdc. Note that the filtering action is just beginning to take place. Connecting a 1000 µF capacitor across the load further drops the ripple factor to 0.096 (Fig. 11-12C); the voltage rises to 11.8 volts. Finally, a 6800 µF capacitor was connected across the load. The ripple factor dropped to 0.014 (Fig. 11-12D). This filtered DC is nearly pure, but still contains a small ripple factor.

Next, the sensitivity of the 'scope was increased from 5 V/cm to 0.1 V/cm (a 50× increase) and the same 6800 µF capacitor was used in parallel with

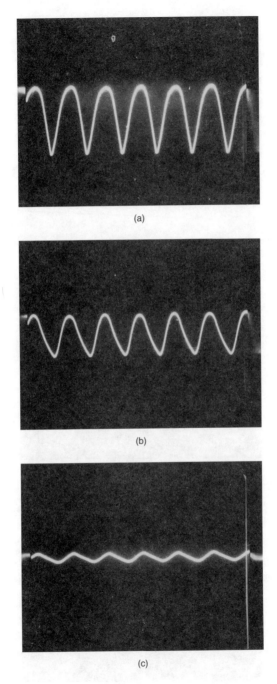

(a)

(b)

(c)

FIGURE 11-12   Ripple factor with different values of filter capacitor.

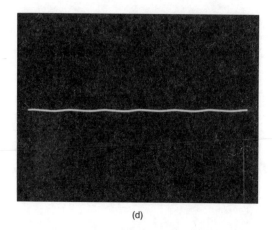

(d)

FIGURE 11-12   (*continued*)

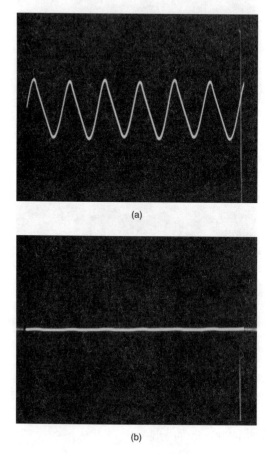

(a)

(b)

FIGURE 11-13   (A) Ripple at input to voltage regulator; (B) Ripple at output of regulator.

the 25 ohm load. The ripple shows up a lot better under this condition (Fig. 11-13A). Recall from above that this waveform represents a ripple factor $r = 0.014$.

A voltage regulator tends to smooth out the ripple considerably. Power supply makers sometimes claim in advertising that their products have the 'equivalent of one farad of output capacitance'. What they mean is that the ripple is reduced by a voltage regulator to the same extent as a $1\,000\,000$ μF capacitor across the load. Want proof? Examine Fig. 11-13B. Using the same 0.1 V/cm 'scope sensitivity and the same power supply, this waveform shows a remarkable drop of ripple compared with Fig. 11-13A. The lesson to learn here is to use a voltage regulator in circuits where it is necessary to reduce the ripple to a very low value.

11-4.1 Pi-network filter circuits

Another form of filter circuit is the RC pi-network shown in Fig. 11-14. Output voltage $V1$ represents a circuit like those above, but output voltage $V2$ has a lower voltage and substantially lower ripple factor. For fullwave circuits the ripple factor is:

$$r = \frac{2.5 \times 10^{-6}}{C1C2R1R_{\mathrm{L}}} \tag{11-9}$$

11-4.2 Working voltage

The other significant rating for filter capacitors, besides capacitance, is the DC working voltage (WVDC). This rating specifies the maximum voltage that the capacitor can sustain safely on a continuous basis (not a transient peak voltage). Because AC line voltages can vary ±15%, and the WVDC rating tolerance may have a ±20% tolerance, it is prudent to use a capacitor with a WVDC rating that is at least half again higher than the normal output

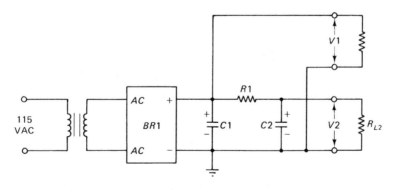

FIGURE 11-14 Two-stage RC ripple filter circuit.

voltage expected. That is, the minimum WVDC rating of the capacitor should be 1.5 times the maximum output voltage of the rectifier. If a filter capacitor has a WVDC rating close to the power supply output voltage, then a short life and a spectacular end may be the fate of the power supply.

11-5 VOLTAGE REGULATION

The output voltage of ordinary rectifier/filter DC power supplies is not stable, but rather, it may vary considerably over time. There are two main sources of variation in the output of this type of power supply. First, there is always a certain fluctuation of the AC input voltage. Ordinary commercial power lines vary from 105 to 120 volts AC (rms) normally, and may droop to less than 100 volts during power 'brownouts'.

The second source of variation is created by load variation (see Fig. 11-15). The problem is caused by the fact that real DC power supplies are not ideal. The ideal textbook power supply has zero ohms internal resistance, while real power supplies have a certain amount of internal resistance (represented by R_s in Fig. 11-15). When current is drawn from the power supply there is a voltage drop ($V1$) across the internal resistance, and this voltage is subtracted from the available voltage (V). In an ideal power supply, output voltage V_o is the same as V, but in real supplies V_o is equal to ($V - V1$). Because $V1$ varies with changes in the load current I_o, the output voltage will also vary with changes in current demand.

The 'goodness' or 'badness' of a power supply can be defined in terms of its percentage regulation. This specification is a measure of how badly the voltage changes under changes of load current, and is found from:

$$\%\text{reg} = \frac{(V - V_o)(100\%)}{V} \tag{11-10}$$

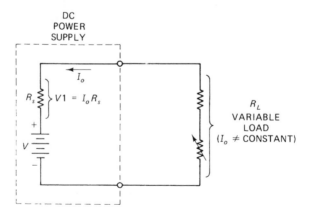

FIGURE 11-15 Source of varying output voltage as load current changes is the voltage divider formed by the load and internal resistance of power supply.

where:

    V is the open-terminal (no output current) output voltage

    V_o is the output voltage under full load current

    %reg is the percentage regulation

Many electronic circuits do not work properly under varying supply voltage conditions. Oscillators and some waveform generators, for example, tend to change frequency if the DC power supply voltage changes. Obviously, some means must be provided to stabilize the DC voltage. The zener diode is perhaps the simplest such voltage regulator device.

11-6 ZENER DIODES

The zener ('zen-ner') diode is a special case of the PN junction diode; Figure 11-16 shows both the circuit symbol (Fig. 11-16A) and I versus V curve (Fig. 11-16B) for a zener diode.

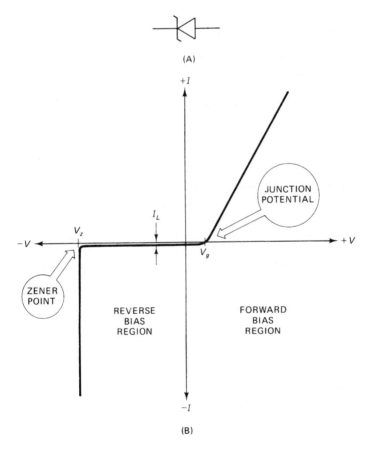

FIGURE 11-16   Zener diode: (A) circuit symbol; (B) I versus V curve.

In the forward bias region operation of the zener diode is the same as for other PN junction diodes. For this case, the anode is positive with respect to the cathode, so a forward bias current $(+I)$ flows. For voltages greater than V_g, which is approximately 0.6 volts to 0.7 volts, the current flow increases approximately linearly with increasing voltage. At potentials less than V_g, the current increases from a small reverse leakage current (I_L) at $V = 0$, to a small forward current at $+V_g$. The diode is, like all other PN junction diodes, nonlinear in this low-voltage region.

The zener diode also acts like any other PN junction diode in the reverse bias region between $V = 0$ volts and the zener potential $-V_z$. In this region, only the small reverse leakage current flows.

At an applied potential of $-V_z$, or greater, the zener diode breaks down and allows a large reverse current to flow. Note in Fig. 11-16B that further increase in $-V$ does not cause an increased voltage drop across the diode. Thus, the zener diode regulates the voltage to its zener potential by clamping action.

11-6.1 Zener diode regulator circuits

A zener diode operates as a parallel or 'shunt' regulator because it is connected in parallel with the load. It regulates by clamping the output voltage across the load to the zener potential. Figure 11-17 shows a typical zener diode regulator circuit that takes advantage of these attributes.

In Fig. 11-17, resistor R_L represents the load placed across the power supply, i.e. the circuits that draw current from the supply. The value of the load resistance is V_z/I_o. Resistor $R1$ is used as a series current limiter to protect the diode. Refer back to Fig. 11-16B to see why it is needed. Note that $-I$ increases sharply when $-V$ reaches the zener potential. If $R1$ is not used, then this current will destroy the diode. One of the tasks in designing a zener diode voltage regulator circuit is selecting the resistance value and power rating of resistor $R1$.

Capacitor $C2$ is optional, and is used to suppress the hash noise generated by the zener diode. The zener process is an avalanche phenomenon, so is

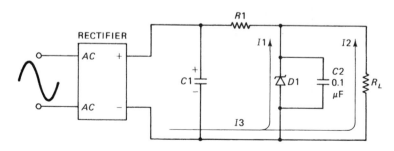

FIGURE 11-17   Zener diode voltage regulator circuit.

inherently noisy. In fact, certain RF and audio noise generators use a zener diode to create the noise signal.

The other capacitor in the circuit, $C1$, is the regular filter capacitor used in any rectifier/filter DC power supply. Its purpose is to smooth the pulsating DC into nearly pure DC. It doesn't really serve a function in the zener regulator circuit, except that the power supply should be filtered prior to the regulator circuit.

The main current drawn from the rectifier ($I3$) is broken into two branches: $I1$ flows through the zener diode and $I2$ flows through the load; $I1$ usually is approximately 10% of $I2$. According to Kirchhoff's law, the relationship between currents is: $I3 = I1 + I2$.

11-6.2   Designing zener diode voltage regulator circuits

When designing zener diode regulators it is necessary to know certain circuit conditions, and then use them to specify (a) the resistance of $R1$; (b) the wattage rating of $R1$; and (c) the wattage rating of zener diode $D1$.

There are three circuit conditions, which are designated I, II and III. The properties of these three conditions are as follows:

Condition I. Variable supply voltage, with constant load current;
Condition II. Constant supply voltage, with variable load current;
Condition III. Variable supply voltage, with load current also variable.

Table 11-1 shows the design equations for all three conditions. Note that the power dissipation expressions for $R1$ and $D1$ are the same for all three

---

TABLE 11-1

Condition I. Variable V_{in}, constant I_o:

$$R1 = \frac{V_{in(max)} - V_z}{1.1 I2}$$

Condition II. Constant V_{in}, variable I_o:

$$R1 = \frac{V_{in(max)} - V_z}{1.1 I2_{max}}$$

Condition III. V_{in} and I_o both variable:

$$R1 = \frac{V_{in(max)} - V_z}{1.1 I2_{max}}$$

Power dissipations for all three conditions:
Diode dissipation:

$$P_{D1} = \frac{[V_{in(max)} - V_z]^2}{R1} - I2 V_z$$

Resistor dissipation:

$$P_{R1} = P_{D1} + I2 V_z$$

---

conditions. Of course, V_{in} and $V_{in(max)}$ are the same for the constant supply voltage case.

A power dissipation of 0.803 watts means that a 1 watt or greater rating is required for the zener diode ($D1$). For best reliability, maintain at least 2:1 ratio between the rated and dissipated wattage. Recommend a rating >1.6 watts (which means 2 watt standard value) diode. For most intermittent or low duty-cycle applications, however, a 1 watt zener diode should suffice.

The required power rating of $R1$ is found from calculating the actual power dissipation of $R1$, and then picking a higher standard wattage rating. The power dissipation of $R1$ is:

$$P_{R1} = P_{d1} + (I2 \times V_z)$$

$$P_{R1} = (0.803 \text{ watts}) + ((0.012)(6.8))$$

$$P_{R1} = 0.803 + 0.816 \text{ watts} = 1.62 \text{ watts}$$

Since P_{R1} is 1.62 watts, use at least 2 watts as the rating of $R1$; a 3 watt or 5 watt resistor would yield even greater reliability.

The component values for this example are 16.7 ohms at 2 watts (or more) for $R1$, and 6.8 volts at 1 watt (or more) for $D1$.

11-7 INCREASING VOLTAGE REGULATOR OUTPUT CURRENT

The output current that can be supplied by a zener diode voltage regulator is somewhat limited. In cases were a larger output current is needed, it is possible to amplify the effect of the zener diode by using it to control the base of a series-pass transistor ($Q1$ in Fig. 11-18). The output voltage produced by this circuit is approximately 0.6 to 0.7 volts less than the zener potential.

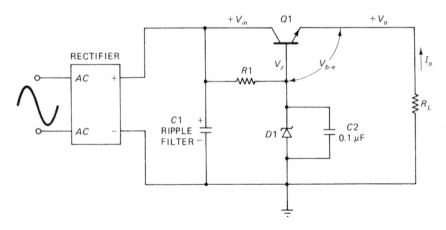

FIGURE 11-18   Series-pass voltage regulator circuit.

This reduction is accounted for by the base–emitter potential of the transistor, V_{b-e}. The output current rating is limited by the collector current rating of the transistor, with due regard for the collector dissipation. The collector will dissipate a power of $(V_{in} - V_o) + I_o$. If there is a large difference between V_{in} and V_o, then it is possible to exceed the maximum collector dissipation rating of $Q1$ even if less than the maximum collector current flows in the circuit. The output current must be limited to a value less than that required to exceed the collector dissipation under the voltage difference conditions established in the circuit.

Another series-pass voltage regulator circuit is the feedback regulator shown in Fig. 11-19. In this circuit, a sample of the output voltage and a reference potential are applied to the differential inputs of a feedback amplifier $(A1)$. When the difference between V_A and V_{ref} is non-zero, the amplifier drives the base of transistor $Q1$ harder, thereby increasing the output voltage. The actual voltage will be stable at a point determined by V_{ref}.

The circuit in Fig. 11-19 shows a feature that is highly useful in high current DC power supplies, especially where the power supply must be operated more than a few inches from the load. The voltage divider $R1/R2$ takes the sample of output voltage V_o that drives $A1$. The lines from the positive output and negative output to the voltage divider are separate from the main current-carrying lines. This arrangement makes it possible to place these 'sense' lines at the points in the actual circuit where the value of V_o must be maintained at a precise value. For example, in a microcomputer that uses high current TTL devices, it matters little that $+5$ Vdc is maintained at the output of the DC power supply: it matters a lot, however, that the voltage at

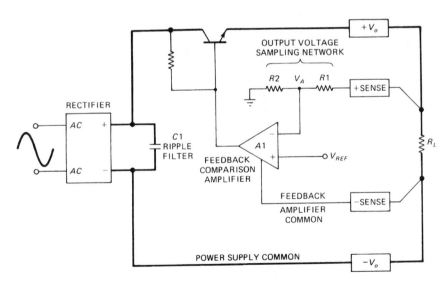

FIGURE 11-19   Sense amplifier series-pass voltage regulator circuit.

the microcomputer printed circuit board is +5 VDC. If the (+)SENSE line is connected to the +5 VDC bus of the computer, and the (−)SENSE line is connected to the ground bus, then the feedback power supply will keep the voltage at the rated value at the PCB, not at the power supply. This method 'servos out' $I \times R$ drop in the power supply lines.

11-8 THREE-TERMINAL IC VOLTAGE REGULATORS _____

Voltage regulators for low current levels (up to 5 amperes) are reasonably simple to build now that simple three-terminal IC regulators are available. The circuit used with positive three-terminal regulators is shown in Fig. 11-20, while typical package styles are shown in Fig. 11-21. Capacitor $C1$ is the normal ripple filter capacitor, and should have a value of 1000 μF per ampere of load current (some authorities insist on 2000 μF/ampere). Capacitor $C4$ is used to improve the transient response to sudden increases in current demand

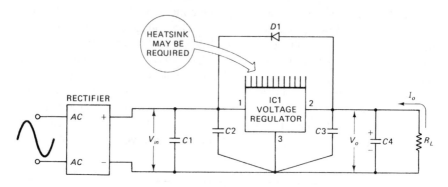

FIGURE 11-20 Three-terminal IC voltage regulator circuit.

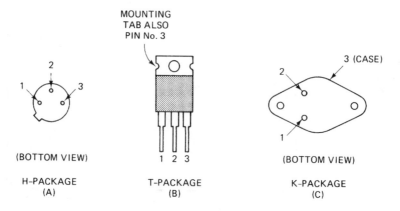

FIGURE 11-21 Three-terminal IC voltage regulator packages.

(something that happens in digital circuits). Capacitor C4 should have a value of approximately 100 μF/ampere load current. Capacitors C2 and C3 are used to improve the immunity of the voltage regulator to transient noise impulses. These capacitors are usually 0.1 μF to 1 μF, and are to be mounted as close as possible to the body of the voltage regulator IC1.

Diode D1 is not shown in a lot of circuits, but is highly recommended for applications where C4 is used. If the diode is not present, then charge stored in C4 would be dumped back into the regulator when the circuit is turned off. That current has been implicated in poor regulator reliability. The mechanism of failure is that the normally reverse biased PN junction formed by the IC regulator substrate and the circuitry become forward biased under these conditions. This situation allows a destructive current to flow. The diode should be a 1 ampere type at power supply currents up to 2 amperes, and larger for larger current levels. For most low-voltage, 1 ampere or less, supplies a 1N400x is sufficient.

Several three-terminal IC voltage regulator packages are shown in Fig. 11-21. The H package (Fig. 11-21A) is used at currents up to 100 mA, the TO-220 T package (Fig. 11-21B) at currents up to 750 mA (or 1000 mA if heatsinked), and the TO-3 K package (Fig. 11-21C) at currents to 1 ampere.

There are two general families of IC regulator. One is designated '78xx', in which the 'xx' is replaced with the fixed output voltage rating. Thus, a '7805' is a 5 volt regulator, while a '7812' is a 12 volt regulator. The 'LM-340y-xx' series is also used. The 'y' is the package style (H, K or T) while the 'xx' is the voltage. Thus, an 'LM-340K-05' is a 1 ampere, 5 volt regulator in a 'similar-to-TO3' type-K package; an 'LM-340T-12' is a 12 volt, 750 mA regulator in a plastic TO-220 power transistor package.

Negative versions of these regulators are available under the '79xx' and 'LM-320y-xx' designations. Figure 11-22 shows the typical circuit symbol. Note that the pinouts on the voltage regulator device are different from those of the positive regulator.

The minimum input voltage to the three-terminal IC voltage regulator is usually 2.5 volts higher than the rated output voltage. Thus, for a +5 volt

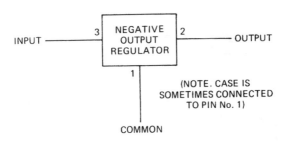

FIGURE 11-22   Negative output three-terminal IC voltage regulator.

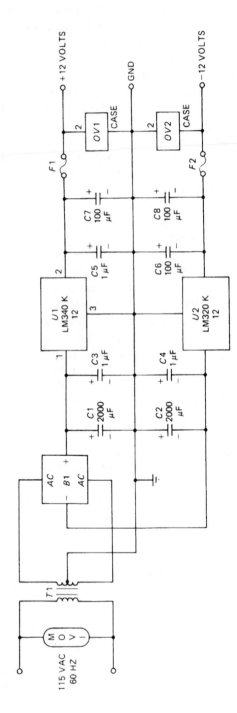

FIGURE 11-23 Dual-polarity regulated DC power supply.

regulator, the minimum allowable input voltage is 7.5 Vdc. The power dissipation is proportional to the voltage difference between this input potential and the output potential. For a 1 ampere regulator, therefore, the dissipation will be 2.5 watts if the minimum voltage is used, and considerably higher if a higher voltage is used. It is recommended that an input voltage that is close to the minimum allowable voltage be used. For +5 volt supplies used in digital projects, a standard 6.3 VAC transformer is sufficient. When full-wave rectified and filtered with 1000 µF/ampere or more, the output voltage will be approximately +8 Vdc. The student is encouraged to work out the arithmetic to prove this statement.

Figure 11-23 shows a dual-polarity DC power supply such as might be used in operational amplifier circuits, some microcomputers and many other applications. The voltage regulator portion of the circuit is a combination of positive and negative output versions of Fig. 11-20. The transformer/rectifier section bears some explanation, however. The rectifier is a 1 ampere bridge stack, but is not used as a regular bridge. The center-tap on the secondary of transformer $T1$ establishes a zero-reference, so the 'bridge' actually consists of a pair of conventional fullwave bridges connected to the same AC source. Thus, the (−) terminal of the bridge stack drives the negative voltage regulator, and the (+) terminal drives the positive voltage regulator. This rectifier is sometimes called a 'halfbridge' rectifier.

11-8.1 Adjustable IC voltage regulators

The IC voltage regulators discussed above are fixed types. They offer an output voltage that is predetermined and unchangeable without extraordinary effort. A variable output voltage regulator, on the other hand, can be programmed to any voltage desired within its range. These devices can be used for either variable DC power supplies, or to supply custom output voltages other than those allowed by the standard fixed voltages. Two similar models are considered here as examples.

The LM-317 and LM-338 are variable DC voltage regulators that are capable of delivering up to 1.5 amperes and 5 amperes, respectively, at voltages up to +32 volts DC. Figure 11-24 shows a typical circuit for these regulators. The input voltage must be 3 volts higher than the maximum output voltage. The output voltage is set by the ratio of two resistors, $R1$ and $R2$, according to the equation:

$$V_o = 1.25 \text{ volts} \left[\frac{R2}{R1} + 1 \right] \tag{11-11}$$

An example from the National Semiconductor, Incorporated *Linear Databook* shows 120 ohms for $R1$, and a 5 kohm potentiometer for $R2$. This combination produces a variable output voltage of 1.2 Vdc to 25 Vdc, when V_{in} is >28 Vdc. Diode $D1$ can be any of the series 1N4002 through 1N4007 for LM-317 supplies, and any 3 ampere type for LM-338 supplies.

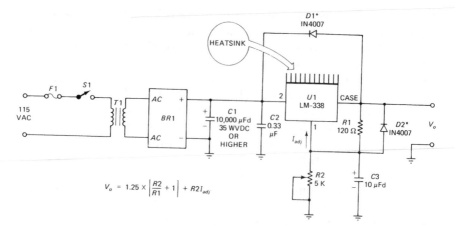

FIGURE 11-24   Variable voltage output regulator circuit based on the LM-335 device.

$$V_o = 1.25 \times \left[\frac{R2}{R1} + 1\right] + R2I_{adj}$$

11-9   OVERVOLTAGE PROTECTION

If the series-pass transistor shorts collector-to-emitter, or if the base control circuit fails, then the input voltage will be applied to the output terminal of the voltage regulator. Because this voltage is often considerably higher than the regulated output voltage, serious damage can result to the electronic circuitry powered by the regulator. A protective circuit called an SCR crowbar provides overvoltage protection to such circuits.

The SCR crowbar (Fig. 11-25) is shunted across the output of the power supply ('V' is the power supply output voltage). A +5 Vdc supply is used as an example. Diode D1 is a zener diode that has a zener potential that is a little higher than the rated output voltage. Zener voltage from 5.6 to 6.8 Vdc can be used to protect a +5 Vdc power supply. Power supplies with other

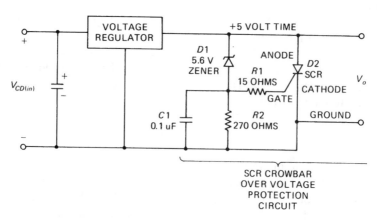

FIGURE 11-25   SCR crowbar overvoltage protection circuit.

output voltages than +5 Vdc can be protected by scaling the zener diode voltage proportionally.

Diode D2 is a silicon controlled rectifier (SCR). This type of diode is open-circuited (i.e. a high resistance in both directions) until a current is caused to flow in the gate terminal. When this occurs, the SCR 'breaks over' and operates as any ordinary PN junction diode.

The gate terminal of diode D2 is controlled by the network around diode D1. When the power supply voltage V exceeds the zener potential of D1, a current is conducted through D1 creating a voltage drop across R2. This voltage drop becomes the source for the gate current that flows in R1 and the gate of D2. At this instant, the SCR becomes conductive and shorts out the power supply line. Either a fuse or fuse resistor can be connected in series with the DC line, and will open-circuit when D2 conducts. The fuse should be in series with the positive input line of the voltage regulator. In a few cases, there is no fuse. In those circuits, the SCR (D2) must have a tremendous current rating because it must carry the short-circuit current of the power supply. It will clamp the output to near ground level until the circuit is turned off.

Lambda Electronics, Inc. offers overvoltage protection integrated circuits. Current levels available are 2, 6, 12, 20 and 35 amperes, with available voltage levels being 5, 6, 12, 15, 18, 20, 24, 28 and 30 volts. Several package styles are used, which also indicate the ampere level. A TO-66 power transistor case is used for 2 amperes, while the TO-3 power transistor case is for 6 amperes. Higher current levels are packaged in epoxy cases. The Lambda overvoltage protection modules are designated with a type number of the form 'L-y-OV-xx', in which 'y' indicates the ampere level, and 'xx' indicates the voltage. Thus, an 'L-6-OV-5' device is a 5 volt, 6 ampere model.

11-10 CURRENT-LIMITING

Another catastrophe that can befall a DC power supply is an output short circuit. For unprotected power supplies, such an event will result in destruction of the circuit. It is possible to place a circuit in the power supply that will provide a 'current knee' above which the supply shuts down. Figure 11-26 shows a representative circuit.

Transistor Q1 in Fig. 11-26 is the series-pass transistor in a regulated power supply, while Q2 is the sense transistor that determines when the current flow is too high. Some IC voltage regulators can also be used with this circuit (and some three-terminal types have the circuit built-in) if they have a 'sense' terminal or some other provision.

Resistor R2 is used to sense the level of current flow. The voltage drop across this resistor provides forward bias to Q2, and is proportional to the current flow:

$$V_{R2} = I_o \times R2 \qquad (11\text{-}12)$$

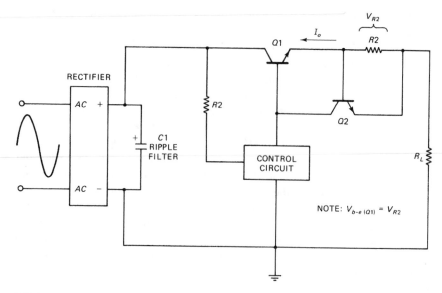

FIGURE 11-26   Current-limiting output circuit.

For silicon transistors, the forward bias voltage required to saturate the transistor is approximately 0.6 volts. When V_{R2} exceeds this critical voltage, transistor $Q2$ is heavily forward biased so its V_{c-e} drops to a very low value, essentially shorting the base–emitter terminals of $Q1$. This actions turns off the power transistor. The value of $R2$ is, therefore:

$$R2 = \frac{0.6 \text{ Vdc}}{I_{o(max)}} \tag{11-13}$$

Consider a practical example. Suppose a computer power supply delivers 10 amperes maximum. The value of $R2$ would be:

$$R2 = 0.6 \text{ Vdc}/I_{o(max)}$$
$$R2 = 0.6 \text{ Vdc}/10 \text{ amperes}$$
$$R2 = 0.06 \text{ ohms}$$

A value of 0.06 ohms (60 milliohms) seems somewhat difficult to achieve, but such a resistor can be made from fine wire. Alternatively, several wire-wound power resistors or fuse resistors can be connected in parallel to form the low value required. For example, a 0.33 ohm resistor is often used as a 'fusistor' in auto radios or the emitter resistor in audio power amplifiers. Five of these resistors in parallel produce very nearly 60 milliohms. Various values of fusistors are available from 0.09 to 1.5 ohms, and these can be paralleled in assorted combinations to produce the required resistance.

11-11 HIGH VOLTAGE TRANSIENT PROTECTION

Experts warn that 20 μs to 500 μs transient pulses of 1500 volts or more strike residential and small business power lines several times per day. In industrial facilities that number may be considerably greater because of the heavy electrical machinery that is often in use. Until digital electronics devices, including computers, were widespread, however, this fact was interesting but somewhat trivial. But high voltage transient pulses can seriously disrupt digital circuits. The circuit may simply fail to operate correctly, or, can be damaged by the transient pulse. If a computer seems to occasionally 'bomb out' while executing a program that ran properly only a short while ago, then suspect these transient pulses as the root cause.

Figure 11-27 shows metal oxide varistor (MOV) devices shunted across the power lines. Normally, only $MOV1$ will be needed, but $MOV2$ and $MOV3$ are recommended in serious cases. These devices are made by General Electric (similar devices are made by others), and can be modeled as a pair of back-to-back zener diodes with a V_z rating of about 180 volts. The purpose of the MOV devices is to clip transient pulses over 180 volts.

Some applications require an LC low-pass filter on the AC power lines. In some very severe transient cases where the MOV is not sufficient, or, in cases in which a strong RF field is present (as in a radio transmitter), or, in cases where the digital device creates RFI, then the LC filter of Fig. 11-28

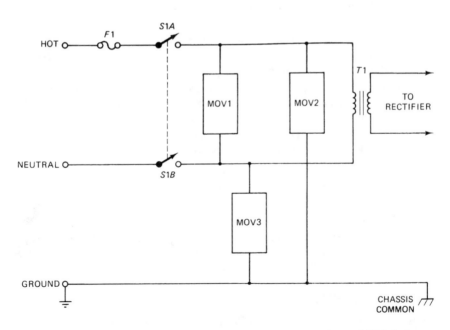

FIGURE 11-27   Input spike suppression with metal oxide semiconductor (MOV) devices.

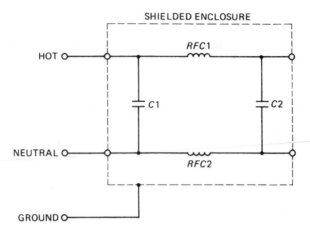

FIGURE 11-28   Power mains EMI filter.

might be indicated. This filter should be mounted as close as possible to the point where the AC power cord enters the equipment.

Several manufacturers offer RFI filters that are shielded, and especially suited for this service. Some models are molded inside of a chassis mounted AC receptacle.

11-12   SUMMARY

1. The DC power supply consists of a transformer to scale voltage levels, a rectifier to convert bidirectional AC to unidirectional pulsating DC, a ripple filter to smooth pulsating DC to nearly pure DC, and (in some circuits) a voltage regulator to stabilize the output voltage.

2. The rectifier is a PN junction diode. It should have a peak inverse voltage rating of >2.83 times the applied AC (RMS) voltage, and a forward current rating high enough to handle the full required load current.

3. The ripple filter may be a single capacitor shunted across the load, or an *RC* pi-network. The filter smooths the ripple to nearly pure DC.

4. The voltage regulator is used to stabilize the output voltage in spite of fluctuation of the input AC voltage and changes in the load current requirements.

5. An overvoltage protection circuit turns off the DC power supply in the event a fault occurs that produces too high an output voltage level.

6. A current-limiting circuit provides protection against either too high an output current demand, or an output short circuit. When the output current increases above a certain point the power supply shuts down.

11-13   RECAPITULATION

Now return to the objectives and Pre-quiz questions at the beginning of the chapter and see how well you can answer them. If you cannot answer certain questions, place a check mark to each and review the appropriate parts of the text. Next, try to answer the questions and work the problems below, using the same procedure.

11-14 STUDENT EXERCISES

WARNING. Some of the exercises in this section involve working with AC power from the mains. This type of power is extremely dangerous if not handled properly. Ideally, you should be working on a bench equipped with isolation transformers. Do not under any circumstances touch exposed AC power line wires, and don't even think about working on the exercises 'hot' (i.e. with power applied). Ask your instructor for information on safely working with AC power circuits.

1. Use a low-voltage AC power transformer (e.g. 12.6 VAC) rated at a forward current of at least 500 mA, and a silicon rectifier diode (e.g. 1N4002 through 1N4007) to produce a halfwave rectifier circuit. Connect a 100 ohm, 5 watt (or more) resistor across the output of the rectifier. Measure the DC output voltage across the resistor, and compare with the calculated values for peak and RMS voltage.

2. Perform the exercise above, and then repeat the experiment using 100 μF/50 WVDC, 1000 μF/50 WVDC, and 10 000 μF/50 WVDC filter capacitors. Compare the results.

3. Build the circuit of a DC power supply (see two previous exercises). Add a 12 volt three-terminal IC voltage regulator. Measure the output voltage, output current, and input voltage to the regulator.

4. Build a negative output voltage regulated DC power supply.

5. Design and build a ±12 Vdc/1 ampere bipolar DC power supply with a common shared between the two polarities.

11-15 QUESTIONS AND PROBLEMS

1. What is the principal job of a rectifier? A ripple filter?

2. Draw the circuits of (a) halfwave rectifier, (b) fullwave rectifier, (c) fullwave bridge rectifier.

3. A rectifier is connected across a load resistor and a shunt capacitor that serves as a ripple filter. What is the minimum peak inverse voltage rating if the transformer delivers 25.6 VAC RMS to the rectifier?

4. List the key specifications for a rectifier diode.

5. State the maximum operating current and output potential of an LM-340K-15 three-terminal IC voltage regulator.

6. A zener diode regulator must regulate a DC supply to +5.6 Vdc. The input DC source varies from 8 to 10 volts. Assume a constant load current of 60 mA. Find the value of series resistor, its power dissipation, and the power dissipation of the zener diode.

7. A single capacitor ripple filter in a 12 Vdc power supply is required to reduce the fullwave ripple to 0.9 when the load current is 10 amperes. Calculate the value of filter capacitance needed.

8. Calculate the ripple factor in a 50/60 Hz halfwave rectifier circuit if a 12.6 VAC (RMS) transformer is used to supply a current of 0.800 amperes; the filter capacitor is 2200 μF.

9. A 12 Vdc power supply must deliver 500 mA approximately 50% of the time. What package should be specified for a three-terminal IC regulator that will do this job?

10. A 270 Vdc power supply is filtered by a 60 μF/350 WVDC capacitor. The voltage can vary ±15%, and the filter capacitor has a ±20% tolerance in the WVDC rating. Calculate the 'worst case' applied voltage and WVDC rating to determine if the capacitor selection was appropriate.

11. A DC power supply delivers 10 Vdc at 250 mA to a resistive load. Calculate the ripple factor of the output from the pi-section RC filter if $R1 = 50$ ohms, $C1 = 2000$ μF and $C2 = 220$ μF.

12. Find the percentage regulation if a DC power supply output voltage drops from 14 to 10 volts as the output current is raised from zero to 1.5 amperes (which is the maximum allowable output current for this power supply).

13. A 9.1 VDC zener diode must regulate in the presence of a supply voltage that varies from +16 to +22 volts. The current remains constant at 140 milliamperes (0.14 amperes). Calculate the value of current-limiting resistor $R1$, the power dissipation of $R1$ and the power dissipation of $D1$. Recommend minimum values for the power ratings of $R1$ and $D1$.

14. An LM-338 variable three-terminal IC voltage regulator is connected with the following resistor values: $R1 = 120$ ohms and $R2 = 3300$ ohms. Find (a) the output voltage; (b) the minimum allowable input voltage.

15. A current limiting circuit is being designed for a 22 ampere DC power supply. Calculate the value of series resistor needed to sense the current level.

Index

Output voltage limiting, 263
Output waveform, fullwave, 181

Package type, IC, 9
Package, MIC, 223
Parallel feedback, 219
pCO2, 167
Peak follower, 259ff.
Peak-to-peak amplitude, 279
Periodic waveform, 239
Permitivity, 234
pH, 167
Phase sensitive detector, see PSD
Phase shift, 111, 127, 278
Phase shift oscillator, 277
Philbrick, George, 2
Photoresistor, 181
Phototransistor, 181
Pink noise, 120
PN junction, 45
pO2, 167
Position, 165
Post amplifier, 164
Power amplification, 85
Power dissipation, 44
Power divider, 231
Power lines, AC, 177
Power splitter, 231
Power supply rejection ratio, see
    PSRR
Power supply sensitivity (PSS), 108
Power supply sensitivity, 43
Power supply terminal, 45
Power supply voltage variation, 277
Power supply, V+, 46
Power supply, V−, 47
Power, noise, 121
Preamplifier, ECG, 164, 182
Printed circuit, 233
Problems, op-amp, 105
Propagation feedback, 130
Protecting IC amplifiers, 17
PSRR, 43
PSS, 108

Pulmonary artery, 186
Pulse generator, 243
Pulse response, 78
Pulse stretcher, 243
Pulse width controller, 57

Quadrature condition, 286
Quasi-stable state, 242–3, 245

Ramp generator, 267
RC charging curve, 267
RC network, 279
RC phase shift network, 126
RC phase shift oscillator, 277
RC-based waveform generator, 242
Reference current, 206
Reference signal, 180
Refractory period, 248
Refractory state, 248
Regeneration, 239
Relaxation oscillator, 239
Resistor, compensation, 116
Resistor–capacitor network, 127
Response, pulse, 78
Response, square wave, 78
Retriggerable MMV, 249
RF choke, 131
RF interference, 128
RF signal power, 227
Root sum squares, 108
RSS, 108

Sample and hold circuit, see S/H
Saturation, 105
Sawtooth generator, 266, 271
Scales of integration, 12
Schottky noise, 120
Seebeck effect, 118
Sensing circuits, 252
Sensor, viii
Sensor isolation, 188
Separation, channel, 43